The Cessna 172

2nd Edition

The Cessna 172
2nd Edition

Bill Clarke

TAB Books
Division of McGraw-Hill, Inc.
New York San Francisco Washington, D.C. Auckland Bogotá
Caracas Lisbon London Madrid Mexico City Milan
Montreal New Delhi San Juan Singapore
Sydney Tokyo Toronto

pbk   4 5 6 7 8 9 10 11 12   FGR/FGR   9 9 8 7 6

Library of Congress Cataloging-in-Publication Data

Clarke, Bill (Charles W.)
    The Cessna 172/ by Bill Clarke.—2nd ed.
       p.   cm.
    Includes index.
    ISBN 0-8306-4294-3 (pbk.)
    1. Cessna 172 (Private planes) 2. Cessna Aircraft Company.
  I. Title.
  TL686.C4C56 1993
  629.133'343—dc 20
                                      93-13234
                                      CIP

Acquisitions Editor: Jeff Worsinger
Production: Katherine G. Brown
Book Design: Jaclyn J. Boone
Cover Design: Graphics Plus, Hanover, Pa.
                                      4322
                                      AV1

To my wife
Without her keeping me on target,
This book would not exist

Contents

3 Engines   79

4 Airworthiness directives   97

5 Selecting a 172   103

6   Inspections   124

7   Caring for a 172   131

8   Preventive maintenance   143

Appendices

Acknowledgments

THE CESSNA 172—2nd Edition was made possible by the kind assistance and contributions of:

Dean Humphrey and Ron Chapman of Cessna Aircraft Company

David Sakrison, editor of Cessna Owner Magazine

Cessna Owner Organizations

Federal Aviation Administration

Randolph Products Company

Smithsonian Institution

Plus all those other wonderful "airplane people" who provided me with photographs, descriptions, advice, hardware, friendship, and encouragement.

A special thanks to the research librarians at the Voorheesville Public Library, of Voorheesville, New York—the finest small-town library a writer could ask for.

Introduction

CESSNA 172 AIRPLANES ARE THE MOST POPULAR and most affordable of all the four-place airplanes ever produced. This guide will assist the pilot, the owner, and the would-be-owner in gaining a complete understanding of these airplanes. You will learn all about the various models and their differences. Read about the chronic problems, and how to fix them. Learn about modifications that can be made to improve the 172's performance and comfort.

If you're thinking of purchasing a used 172, you will discover where and how to locate a good used one—and, more importantly, how to keep from getting stung in the process. Although the prospective buyer may have a basic idea of what the advertised airplane looks like, a source to review for further information about the airplane, its equipment, and its value, should be available. This book is such a source. A price guide, based on the current used airplane market, is provided at the back of the book. A walk-through of all the purchase paperwork will be detailed, with examples of the forms shown.

Read how to care for your plane, and learn what preventive maintenance you may legally perform yourself. See what an annual inspection is all about. There is even a chapter about float flying and the new avenues of adventure to be found on the water. An avionics section is included to aid you in making practical and economical decisions when you decide to upgrade your avionics.

Hangar-fly the 172 series and see what their pilots have to say. Hear from the mechanics who service them, and read what the National Transportation Safety Board has to say about the 172 and other small family size airplanes.

In this second edition of the book, the recent history of Cessna will be described, modifications to the 172s are emphasized, prices become current, and new aids are provided for pre-purchase inspections.

Ideally, this guide shall help the Cessna 172 owner and pilot by providing as much background material as possible in one reference guide.

The 172, Cessna's most popular production airplane.

1

History of Cessna

ON A JUNE DAY IN 1911, a 31-year-old farmer and mechanic from Rago, Kansas, cranked up the Elbridge engine of his homemade wood-and-fabric airplane and made his first short yet successful flight. Clyde Cessna became the first person to build and fly an airplane west of the Mississippi River and east of the Rockies, laying the cornerstone of today's Cessna Aircraft Company—the world leader in general aviation production and sales (Fig. 1-1).

Thereafter, until America's entry into World War I curtailed civilian flying, Clyde Cessna designed and built one airplane every year and, with firm faith in his own designs, flew them at exhibitions ranging over wide geographic areas. Each year he improved and refined his basic design. But he built no airplanes for sale. Barnstorming was more profitable than airplane sales and certainly more fun (Fig. 1-2).

In the winter of 1916-17, Cessna accepted an invitation from the Jones Motor Car Company to build his newest airplane in their plant in Wichita, Kansas; thus, he pioneered the manufacture of powered aircraft in Wichita—starting that city on its road to fame as the air capital of the world.

On July 5, 1917, he set a notable speed record of 124.62 mph on a cross-country flight from Blackwell, Oklahoma, to Wichita. This record was only the first of many racing and competition triumphs to be scored by Cessna airplanes.

In 1925, with a total of six successful airplane designs to his credit, Cessna joined Walter Beech and Lloyd Stearman in establishing the Travel Air Manufacturing Company in Wichita. Cessna became the company's president. He remained with Travel Air until he sold out to Beech in 1927. Cessna built his first production model airplane in 1927, the four-place full-cantilever high-winged Comet monoplane. Cessna Aircraft Company came into being December 31, 1927.

EARLY CESSNA MILESTONES

1928. Cessna goes into production with the A series, the first full cantilever-wing airplane to enter production in this country (Fig. 1-3). A Cessna AW wins Class A Transcontinental Air Derby from New York to Los Angeles.

1929. Cessna builds a new factory southeast of Wichita on 80 acres of land.

1930. Cessna builds the CG-2 glider to offset the "depression market" sag.

Fig. 1-1. Clyde Cessna (right) and his nephew Dwane Wallace. Dwane became president of Cessna in 1936.

Fig. 1-2. C.V. Cessna and his aeroplane at Burdett, Kansas, in 1914.

Fig. 1-3. An A Series airplane in approximately 1928.

1931. A Cessna AW wins the *Detroit News* Trophy Race and World's Most Efficient Airplane award.

1933. Cessna builds its first retractable landing gear airplane, the CR-2 Racer. Also, the Cessna CR-3 Racer sets a world speed record for airplanes with engines having fewer than 500 cubic inches of displacement—242.35 mph at the 1933 American Air Races.

1935–36. The Cessna Model C-34 wins the *Detroit News* Trophy Race and World's Most Efficient Airplane award in 1935 and 1936, then takes permanent possession of the trophy (Fig. 1-4).

Fig. 1-4. Dwane Wallace used an airplane such as this C-34 for air racing, often using his winnings to meet the Cessna payroll.

1937. Cessna starts producing the Airmaster C-37.

1938–40. Cessna manufactures the Models C-145 and C-165 Airmasters.

1940–45. Cessna builds more than 5400 Bobcats (Model T-50) for World War II use. Cessna's first twin-engine airplane is the first use of a low-wing design by the company (Fig. 1-5).

Fig. 1-5. The T-50 is sometimes called the Bamboo Bomber. One famous T-50 was *Song Bird* in the 1950s TV series Sky King.

1942. Cessna builds a new plant in Hutchinson, Kansas and 750 Waco CG-4A gliders are built.

1946. Cessna returns to commercial production with the Models 120 and 140, introducing spring landing gear with these models, and converts production facilities from welded steel tubing and wood techniques to all-metal methods (Fig. 1-6). Cessna diversifies aircraft production with industrial hydraulic components.

1947. Cessna builds furniture in the Hutchinson plant. Production begins on the five-place Models 190 and 195, Cessna's first all-metal airplanes.

1948. Cessna enters the four-place airplane market with the Model 170.

Fig. 1-6. Cessna's 120/140 airplanes were the first of the postwar models.

THE 170

The Cessna four-place line of airplanes started in 1948 with the introduction of the Model 170 (Fig. 1-7 and Fig. 1-8). The first 170s had a metal fuselage, fabric-covered wings, and two wing struts on each side, and conventional landing gear, as did most airplanes of that era. The 170 was powered with a Continental C-145-2 engine.

Cessna Aircraft Company

Fig. 1-7. This Cessna 170B was built in the early 1950s. Although it marked the end of the 170 line, it was just as modern as today.

The 170 was updated to the 170A model in 1949; dual wing struts and fabric-covered wings were out and all metal wings were in. There was also a slight change in the tail fin. The same Continental engine was standard; the Franklin O-300 was an option. An 18-gallon auxiliary fuel tank was also optional. The A Model was in production for three years.

The last 170 model was the 170B, introduced in 1952. The 170B came with the Continental C-145-2 engine or the Franklin 165 and had the large Para-Lift flaps that are found on Cessna airplanes.

By 1957, the year after the Model 172 was introduced, the demand for the "easy drive-it" airplane saw sales of the conventional-geared 170 lagging and sales of the tri-geared 172 rising. The original 172 was a 170 with a nosewheel and a different rudder.

Production halted after building 5136 Model 170 airplanes. Although production of the 170 ended nearly 30 years ago, they are still popular and command high prices when sold.

Model 170 specifications

Engine
    Make: Continental
    Model: C-145-2 (or Franklin option)
(Continued on p. 6.)

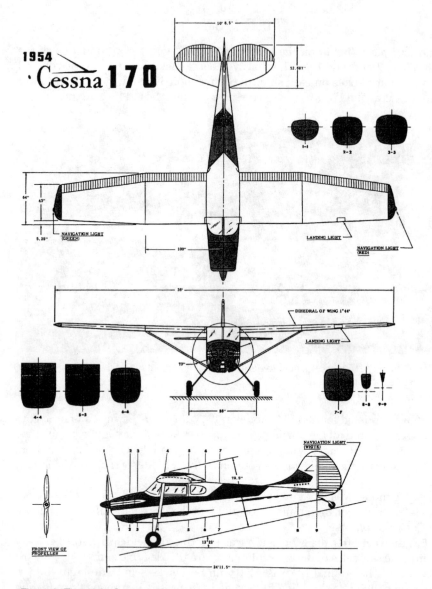

Fig. 1-8. The 1954 Cessna 170. Cessna Aircraft Company

(Continued from p. 5.)

    hp: 145
    TBO: 1800
Seats: 4
Speed
    Max: 140 mph

(Continued on p. 7.)

Cruise: 120 mph
Stall: 52 mph
Fuel Capacity: 42 gal
Rate of Climb: 690 fpm
Transitions
    Takeoff over 50-ft Obstacle: 1820 ft
    Ground run: 700 ft
    Landing over 50-ft Obstacle: 1145 ft
    Ground roll: 500 ft
Weights
    Gross: 2200 lbs
    Empty: 1260 lbs
Dimensions
    Length: 25 ft 0 in
    Height: 6 ft 5 in
    Span: 36 ft 0 in

SUPPORT FOR THE 170

The International Cessna 170 Association was officially formed in order to keep the 170 flying as inexpensively and as easily as possible. The organization's purpose is to furnish information about service, parts, and flying techniques to its members. In addition, general aviation gossip, insurance and safety data, as well as other information is exchanged.

The association issues a quarterly magazine, *The 170 News*, that has photos, news items, classified ads, articles, and letters. A newsletter that is published 11 times a year includes additional classified advertisements that are free for members.

Regional get-togethers are supplemented by an annual convention in the summer, which is advertised as a "week of family fun with a little education thrown in." For further information about the International Cessna 170 Association, contact:

International Cessna 170 Association, Incorporated
P.O. Box 1667, Lebanon, MO 65536
(417) 532-4847

MORE HISTORY

1949. Cessna converts to metal-covered wings on Model 120, 140, and 170 airplanes, ending the fabric-covered airplane era at Cessna. This was the first year of the 170A.

1950. Cessna re-enters military prime contract business with the L-19 Bird Dog. This model incorporates Cessna's first high-lift wing flaps.

1952. Cessna Industrial Hydraulics Division is established at Hutchinson, Kansas. Cessna introduces the Para-Lift wing flaps on commercial airplanes (Model 170B).

1953. The Model 180 enters production. (The 180 will later become famous as a bushplane in remote regions of the world.) Cessna flies its first jet airplane, the XT-37A.

1955. Cessna initiates production of the T-37A (Fig. 1-9).

1956. Cessna introduces the tricycle landing gear with Models 172 and 182 (Fig. 1-10). New sales phrases herald the 172, and will be heard for many years: the Land-O-Matic tricycle landing gear for "driving" the airplane into the air and back onto the ground, plus Para-Lift flaps that originally appeared on the 170B.

Fig. 1-9. The Cessna T-37, used by the U.S. Air Force as a jet trainer.

Fig. 1-10. A 1956 Cessna 172 with many to follow, encompassing four decades.

MODERN HISTORY

1957. Sales lag on the conventional-gear Model 170 and production is halted.

1958. The Cessna Model 175 is introduced, essentially a 172 with a geared, 175-hp engine.

1959. Cessna phases out the T-37A and starts production of the T-37B jet trainer. Cessna re-enters the two-place airplane market with the Model 150. Cessna purchases Aircraft Radio Corporation as a wholly-owned subsidiary.

1960. All Cessna production airplanes adopt swept tails except the Model 150 and Model 180. Cessna purchases 49 percent interest in Reims Aviation, Reims, France.

1961. Skyknight production starts; it is Cessna's first supercharged twin-engine airplane.

1962. Omni-Vision wraparound windshields are introduced on Models 210 and 182.

1963. Omni-Vision is introduced on Model 172s. Cessna produces its 50,000th airplane, a Skyhawk 172.

1964. Cessna is presented the President's E Award for excellence in exporting. Cessna receives a contract for 170 T-41A (Model 172) aircraft for USAF pilot training (Fig. 1-11).

Fig. 1-11. The T-41, Cessna's military primary trainer.

1965. The 10,000th Model 172 is delivered to a flying club in Elaine, Arkansas. Production reaches the milestone of one airplane every 23 minutes during the eight-hour working day (Fig. 1-12 and Fig. 1-13).

1966. Cessna takes a big step to broaden the base of the private aircraft market by launching a worldwide learn-to-fly campaign, increasing production of the 1966 Model 150 two-place trainer to 3000 units and reducing the price of the aircraft by more than 10 percent to make it more readily available. Cessna delivers its 60,000th airplane to an Oklahoma supermarket manager. Argentina approved aircraft manufacturing in that country.

1967. The 75,000th Cessna airplane is delivered. Cessna delivers three versions of the T-41 to the Air Force, the Air Force Academy, and the Army. The first A-37s are delivered, marking the first time a general aviation manufacturer built a combat-designated airplane.

Fig. 1-12. Pawnee Division of Cessna, where the 172s were built.

Fig. 1-13. The assembly line that once produced an airplane every 23 minutes.

1968. The 10,000th Model 150 and the 10,000th Model 182 are delivered. The 1000th T-37 jet trainer is delivered to the U.S. Air Force.

1974. The number of delivered Cessna 172 airplanes surpasses that of the Piper Cubs and Super Cubs at 20,000 (Fig. 1-14).

Fig. 1-14. Pawnee Division's 100,000th airplane was an appropriate model.

1975. Cessna passes the 110,000-airplane mark in total production.

1978. Cessna has pilot training centers in 33 countries worldwide.

1979. Cessna produced 9197 airplanes for the year.

1980. The Silver Anniversary of the Model 172 airplane with more than 31,000 delivered.

1983. Cessna posts the first yearly loss in the company's 55-year history.

FINAL APPROACH

After 25 continuous years of production, 1980 was the Silver Anniversary of the Model 172 airplane and a total exceeding 31,000 Model 172s built; however, by this year storm clouds were forming on the horizon for all of general aviation. Cessna was forced to lay off 750 workers in anticipation of a reduced demand for single-engine airplanes.

Cessna posted the first yearly loss in the company's 55-year history in 1983 and in 1984 experienced zero production on some models. Sales lagged as the selling prices accelerated beyond what prospective purchasers were willing to spend. No 172s were produced in 1987. Increased labor costs, rising materials cost, soaring engine prices, and product liability costs contributed to the decline of Cessna and most general aviation aircraft production.

Textron Incorporated acquired Cessna Aircraft in 1992. The company continues to fully support all of the roughly 130,000 Cessna airplanes that are still in operation around the world, according to a company spokesman. In spite of persistent rumors of new production, Cessna states that due to the product liability problem they have no plans to start building more light aircraft.

THE 172 FAMILY

Initial production of the Model 172 airplanes—and the subsequent improved versions—started a success story unlike any other in modern aviation. Cessna introduced the 172 in 1955. It has been accepted as has no other lightplane based upon the airplane's simplicity and economy, coupled with its abilities.

The 172s provide a means of air transport that is nontaxing on the pilot because there are no complex systems or controls to handle. Likewise, the absence of complex aircraft systems translates to reduced maintenance expenses. The 172's economy includes the engine, which consumes less fuel compared to fuel consumption of complex airplanes. This simplicity and economy was introduced in 1955 to a market that was ready for family airplanes. Perhaps the most important virtue of the 1955 172 was the tricycle landing gear. Also, the 172 was the only all-metal airplane in its class.

Production of four-place conventional-geared airplanes slacked off between 1946 and 1956 with cessation of the Piper Pacer (replaced with the Tri-Pacer, also a tri-geared airplane), the Aeronca Sedan, the Luscombe 11 Sedan, the Stinson Voyager, and the Taylorcraft Ranchwagon. Most of these airplanes were also partially or completely fabric-covered.

Beechcraft Bonanzas and Navions were on the other end of the scale of four-place airplanes. Both airplanes were considerably more complex than the 172; therefore they were expensive to buy and operate and systems were more complicated. Bonanzas and Navions were built for and flown by an entirely different segment of the aviation market.

The 172 is a hard-working, honest airplane capable of providing economical and reliable air transportation. The basic 172 is not completely alone on the Cessna family tree. Cessna tried to capitalize on a good thing, often unsuccessfully, by adding different models of the basic 172 airframe to the product line.

The 175

In 1958, the Model 175 was introduced. Cessna said: "It is an entirely new airplane with new performance figures and a new purpose in life." Other than power, it was really just a 172; the point was proven in 1963, when it reverted to being called a 172 (Fig. 1-15).

The Model 175 had a 175-hp geared Continental GO-300 engine. Although Cessna tried to push these planes, they did not achieve a high level of success and production was halted after only a few years. One reason for nonacceptance was the high (3200) rpm engine operation, reduced to 2400 propeller rpm via a gearbox. The high rpm seemed unnatural, and pilots would not accept it. Additionally, the high speed wore the engine out faster, and the gearboxes had a high—expensive—failure rate.

The 175s are more than 20 years old now and, depending upon the engine situation, can represent good buys. The original geared engine had a low TBO, as you will see in the specifications, and is considerably more expensive to repair and rebuild than the nongeared O-300 engine found in the Model 172. As a result of the engine problems, many owners saw fit to replace the geared engine with a standard powerplant. Properly done, these engines enhance the desirability of the 175; however, no matter what is changed, the airframe is still a Model 175. Cessna began calling it the "172 Powermatic" in 1963, but it was a Model 175 with a new name because specifications were the same as the 1962 Model 175.

Fig. 1-15. 1961 Cessna 175B.

Model 175 specifications

Speed for the Skylark version will be slightly more than the standard 175 by approximately 2 mph. There is a corresponding difference in range. All other performance figures remain the same.

Year: 1958 to 1962
Range and Speed:
    Maximum cruise speed: 139 mph
    Range: 52 Gallons, No Reserve: 595 mi 4.3 hrs
    Maximum Range: 52 Gallons, No Reserve: 720 mi
        (39 percent power at 10,000 ft): 102 mph 7.0 hrs
Rate of Climb at Sea Level: 850 fpm
Service Ceiling: 15,900 ft
Baggage: 120 lbs
Wing Loading (lbs/sq ft): 13.4
Power Loading (lbs/hp): 13.4
Fuel Capacity
    Standard: 52 gal
Engine
    1958 Continental GO-300A
    1959 Continental GO-300A
    1960 Continental GO-300C
    1961 Continental GO-300D
        GO-300C in Standard
    1962 Continental GO-300E
    TBO: 1200 hrs
    Power: 175 hp

Dimensions
    Wingspan: 36 ft 0 in
    Wing Area (sq ft): 175
    Length: 25 ft 0 in
    Height: 8 ft 06 in
Weight
    Gross Weight: 2350 lbs
    Empty Weight
        (Skylark) 1395 lbs
        (Standard) 1325 lbs
    Useful Load
        (Skylark) 955 lbs
        (Standard) 1025 lbs

The 172 XP

Much later in the 172's history, Cessna again introduced a version with greater power; the Hawk XP series was introduced in 1977. The airplane had a Teledyne Continental IO-360 fuel-injected engine of 195 hp. Production lasted only five years, similar to the Skylark. Many claimed little was gained in performance to justify the added initial purchase expense and continued higher operating expense of the larger engine. The XPs did find acceptance as floatplanes and the floatplane pilots and owners say that the extra power is the deciding factor (Fig. 1-16).

Cessna Aircraft Company

Fig. 1-16. 1980 Cessna Hawk XP.

Model R 172 (XP) specifications

Year: 1977 to 1981 (Hawk XP and Hawk XP II)
Speed
    Top Speed at Sea Level: 134 kts
    Cruise, 80 percent power at 5500 ft: 131 kts

Range (with 45 minute reserve)
    Cruise, 80 percent power at 5500 ft: 485 nm
        49 Gallons Usable Fuel: 3.7 hrs
    Cruise, 80 percent power at 6000 ft: 635 nm
        66 Gallons Usable Fuel: 4.9 hrs
    Maximum Range at 10,000 ft: 595 nm
        49 Gallons Usable Fuel: 5.6 hrs
    Maximum Range at 10,000 ft: 815 nm
        66 Gallons Usable Fuel: 8.7 hrs
Rate of Climb at Sea Level: 870 fpm
Service Ceiling: 17,000 ft
Takeoff
    Ground Run: 850 ft
    Over 50 ft Obstacle: 1360 ft
Landing
    Ground Roll: 620 ft
    Over 50-ft Obstacle: 1270 ft
Stall Speed
    Flaps Up, Power Off: 53 kts
    Flaps Down, Power Off: 46 kts
Baggage: 200 lbs
Wing Loading (lbs/sq ft): 14.7
Power Loading (lbs/hp): 13.1
Fuel Capacity
    Standard: 52 gal
    w/Optional tanks: 68 gal
Engine
    1977 Continental IO-360-K
    1978 Continental IO-360-K
    1979 Continental IO-360-KB
    1980 Continental IO-360-KB
    1981 Continental IO-360-KB
    TBO: (1500 hrs on K model) 2000 hrs
    Power: (at 2600 rpm) 195 hp
    Propeller: (diameter) C/S 76 in
Dimensions
    Wingspan: 35 ft 10 in
    Wing Area (sq ft): 174
    Length: 27 ft 02 in
    Height: 8 ft 09 in
Weight
    Maximum Weight: 2550 lbs
    Empty Weight
        (Hawk XP): 1549 lbs
        (Hawk XP II): 1573 lbs
    Useful Load
        (Hawk XP): 1001 lbs
        (Hawk XP II): 977 lbs

The Cutlass RG

The Cutlass RG was introduced in 1980 as a retractable gear version of the 172 with an Avco Lycoming 180-hp O-360, selling for approximately $19,000 more than the basic cost of a Model 172. Although offering better speeds than the standard 172, operating and maintenance costs were predictably higher (Fig. 1-17).

Fig. 1-17. Cutlass RG.

Model 172 Cutlass RG specifications

Year: 1980 to 1985
Speed
    Top Speed at Sea Level: 145 kts
    Cruise, 75 percent power: 140 kts
Range (with 45 minute reserve)
    Cruise 75 percent power at 9000 ft: 720 nm
        62 Gallons Usable Fuel: 5.3 hrs
Maximum Range at 10,000 ft: 840 nm/7.7 hrs
Rate of Climb at Sea Level: 800 fpm
Service Ceiling: 16,800 ft
Takeoff
    Ground Run: 1060 ft
    Over 50-ft Obstacle: 1775 ft
Landing
    Ground Roll: 625 ft
    Over 50-ft Obstacle: 1340 ft
Stall Speed
    Flaps Up, Power Off: 54 kts
    Flaps Down, Power Off: 50 kts
Baggage: 200 lbs
Fuel Capacity
    Standard: 66 gal
Engine
    Make: Lycoming
    Model: O-360-F1A6
    TBO: 2000 hrs

Power: (at 2700 rpm) 180 hp
Propeller: Constant Speed
Dimensions
    Wingspan: 36 ft 01 in
    Wing Area (sq ft): 174
    Length: 26 ft 11 in
    Height: 8 ft 09 in
Weight
    Gross Weight: 2650 lbs
    Empty Weight
        (Cutlass RG): 1615 lbs
        (Cutlass RG II): 1644 lbs
    Useful Load
        (Cutlass RG): 1043 lbs
        (Cutlass RG II): 1014 lbs

The Cutlass Q

The last offering was the nonretractable Cutlass introduced in 1983. The 172 Q was also powered with a 180-hp engine.

Model 172 (Q) Cutlass specifications

Year: 1983 to 1985
Speed
    Top Speed at Sea Level: 124 kts
    Cruise, 75 percent power: 122 kts
Range (with 45 minute reserve)
    Cruise at 75 percent power at 8500 ft: 475 nm
        50 Gallons Usable Fuel: 4.0 hrs
        62 Gallons Usable Fuel: 620 nm/5.2 hrs
Maximum Range at 10,000 ft
        50 Gallons Usable Fuel: 600 nm/6.4 hrs
        62 Gallons Usable Fuel: 775 nm/8.2 hrs
Rate of Climb at Sea Level: 680 fpm
Service Ceiling: 17,000 ft
Takeoff
    Ground Run: 960 ft
    Over 50-ft Obstacle: 1690 ft
Landing
    Ground Roll: 575 ft
    Over 50-ft Obstacle: 1335 ft
Stall Speed
    Flaps Up, Power Off: 53 kts
    Flaps Down, Power Off: 48 kts
Baggage: 120 lbs
Fuel Capacity
    Standard: 54 gal

Long range tanks: 68 gal
Engine
    Make: Lycoming
    Model: O-360-A4N
    TBO: 2000 hrs
    Power (at 2700 rpm): 180 hp
    Propeller: Fixed Pitch
Dimensions
    Wingspan: 36 ft 01 in
    Wing Area (sq ft): 174
    Length: 26 ft 11 in
    Height: 8 ft 09 in
Weight
    Gross Weight: 2550 lbs
    Empty Weight
        (Cutlass): 1480 lbs
        (Cutlass II): 1500 lbs
    Useful Load
        (Cutlass): 1078 lbs
        (Cutlass II): 1058 lbs

172 CHANGES

Cessna typically updated the 172 every year, usually introducing all the new Cessna airplanes to coincide with new automobile announcements. Changes and improvements, except exterior and interior styling, are detailed in the following list.

1955. In November, the Cessna 172 is introduced at a price of $8995 for the 1956 model. Among the features that would make it the "most popular airplane" were the Land-O-Matic gear and the Para-Lift flaps.

1958. The Model 175 is introduced with the more powerful Continental GO-300 engine (Fig. 1-18 and Fig. 1-19). A new world's endurance flight record is set by a 172 remaining aloft for 1200 hours, 16 minutes, and 10 seconds—50 days.

1959. Electric fuel gauges are offered, along with the die-cast wheels and gear-toothed brakes.

1960. The Flight Sweep swept-back vertical tail is added and 172s are made available on floats (Fig. 1-20). The cost of a 1960 Model 172 is $9450.

1961. The Skyhawk name is added to upgraded models that include certain equipment, such as avionics and appearance packages. The landing gear was shortened 3 inches and an improved cowling was introduced.

1962. The 175 is dropped from production. An autopilot is available for the first time. The pull-handle starter is replaced with a starter key button. Wingtips and speed fairings have been redesigned—they will fit all models from 150 through 180. Six-way adjustable seats become available. Family seats are introduced to allow the 172 to carry as many as six passengers (Fig. 1-21). The Model 172 sells for $9,895 and the Skyhawk sells for $11,590.

1963. The Model 172 Powermatic, an updated Model 175, is added to the 172 line. Omni-Vision windows permit 360° visual scanning. The two-piece windshield is replaced with a single-piece unit (Fig. 1-22) and new rudder and brake pedals are in-

Fig. 1-18. Cowling of a 172.

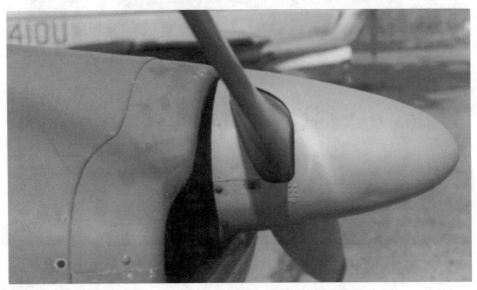

Fig. 1-19. Cowling of a 175. Notice the raised area to house the gearbox.

stalled. The basic 172 sells for $10,245, the 172 Powermatic for $13,275, the Skyhawk for $11,995, and the Skyhawk Powermatic for $14,650.

1964. Manual flaps have been replaced by electric flaps. Push-to-reset circuit breakers are installed on the electrical distribution panel.

1966. The 172 sells for $12,450 and the Skyhawk for $13,300.

Fig. 1-20. Swept-back tail on a "fastback."

Fig. 1-21. It is possible to carry six in the 172—with a weight limitation.

1967. Short-stroke nose gear is added, offering less drag and better appearance. A shock-mounted cowl is installed resulting in lower noise levels in the cabin. A pneumatic stall warning horn and plastic wingtips are standard on all versions. The 172 sells for $10,950 and the Skyhawk for $12,750.

1968. The Blue Streak Lycoming O-320-E2D 150-hp engine replaces the Continental—5 more horsepower and 45 fewer pounds. A new cowling encloses the Lycoming engine and the T primary instrument arrangement was introduced.

Fig. 1-22. This late model shows both the omni-vision windows and the one-piece windshield.

1969. Long-range fuel tanks of 52-gallon capacity are offered as an option. A dorsal fin is installed and the rudder restyled. The rear windows are increased in size by 16 square inches. The 172 sells for $12,500 and the Skyhawk for $13,995.

1970. Conical-camber wingtips are added for better performance (Fig. 1-23 and Fig. 1-24), and fully articulating seats are available, as an option, for passenger comfort.

Fig. 1-23. Original wingtips of the first 172s.

1971. Tubular-strut landing gear replaces the flat-spring gear. Tread width is more than 1 foot wider than the older style (Fig. 1-25 and Fig. 1-26). The 172 sells for $13,425 and the Skyhawk for $14,995.

1972. A new, enlarged, dorsal fin is added (Fig. 1-27 and Fig. 1-28). The cabin doors and baggage door are fabricated with metal-to-metal bonding—no rivets.

1973. A new Camber-lift wing improves low-speed handling characteristics.

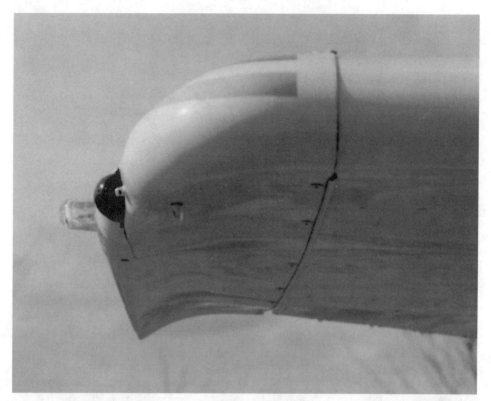

Fig. 1-24. Improved conical tips.

Fig. 1-25. Original flat steel main gear.

Fig. 1-26. Improved tubular steel main gear.

Fig. 1-27. Pre-1972 fin.

Fig. 1-28. Post-1972 fin.

1974. The Skyhawk II package is introduced with a second navcom, an ADF, and a transponder. The baggage compartment is increased in size, and optional dual nose-mounted landing lights are available (Fig. 1-29 and Fig. 1-30).

Fig. 1-29. Single landing light.

Fig. 1-30. Double landing lights.

1975. The 172 sells for $16,055, the Skyhawk for $17,890, and the Skyhawk II for $20,335.

1976. The instrument panel is reorganized to accommodate the ever-increasing amount of avionics available.

1977. The Hawk XP is introduced with a 195-hp fuel-injected engine using a constant-speed prop; it is actually an updated Reims Rocket (built under license in France) and similar to the military T-41 (referred to as the Mescalero by the U.S. Army). The Skyhawk/100 (for 100 octane LL avgas) is introduced. A new engine, the 160-hp Avco Lycoming O-320-H2AD, is installed in the 100s. Rudder trim is available as an option and a preselect flap control is standard. The Skyhawk/100 sells for $22,300, the Skyhawk/100 II for $24,990, the Hawk XP for $29,950, and the Hawk XP II for $32,650.

1978. A 28-volt electric system is installed, and air-conditioning is an option. The Hawk XP is certified for floats.

1979. Flap extension speed is increased to 110 knots on the Skyhawk and Hawk XP. Optional integral fuel cells become available for the Hawk XP, increasing the usable fuel to 66 gallons.

1980. More than 31,000 172s had been built when the airplane celebrated its silver anniversary. A slimmer door post increased visibility. A rounded leading edge is added on the elevator.

1981. A new engine, the Avco Lycoming O-320-D2J, is installed. Maximum flap extension is reduced from 40° to 30° on all models. A wet wing optional fuel tank is available, making a total fuel capacity of 62 gallons usable for the Skyhawk. This is the last year for the Hawk XP. The Skyhawk sells for $33,950, the Skyhawk II for $31,810, the

Skyhawk with the Nav/Pac for $42,460, the Hawk XP for $41,850, the Hawk XP II for $45,935, and the Hawk XP II with Nav/Pac for $50,790.

1982. Landing lights that are more than double the power of previous lights are installed in the leading edge of the wing (Fig. 1-31).

1983. Sound reduction improvements include a thick plexiglass windshield and side windows. Redesigned cabin door latching pins are introduced.

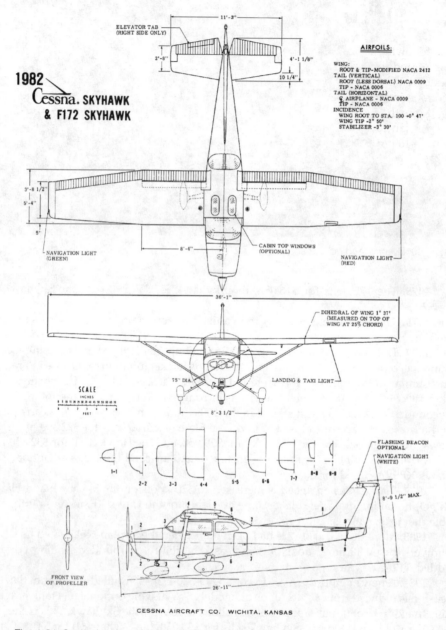

Fig. 1-31. Scale drawing of the 1982 Cessna 172. Cessna Aircraft Company

1984. The Skyhawk sells for $44,000, the Skyhawk II for $48,940, and the Skyhawk II with Nav/Pac for $54,380.

1985. The base price for a Skyhawk is $49,600, which is stripped down with no essential avionics. An average equipped plane is priced at $67,725.

1986. The price for an equipped airplane—$74,705—exceeded what buyers were willing to pay. The Cessna 172 came to the end of the line.

PRODUCTION FIGURES

The following charts provided by Cessna Aircraft Company account for the production periods and quantities of Model 172 and 175 airplanes. Totals are for the U.S.-made versions only; the figures would be higher if 2229 172s produced by Reims Aviation in France were included.

More than 37,000 Cessna 172s were produced (USA version). This is the record for any airplane ever manufactured in the world—even greater than the German Messerschmitt 109 of World War II fame—even without including the 175s or the Reims 172s.

Fig. 1-32. Bolstered by the immense popularity of the 172 series, Cessna built and tested other experimental designs, but few were ever seen in public.

Model 172

Year

1955	173 (includes 1956 models)
1956	1419
1957	939
1958	790
1959	874

1960	1015
1961	903
1962	889
1963	1146
1964	1401
1965	1436
1966	1597
1967	839
1968	1206
1969	1170
1970	759
1971	827
1972	984
1973	1550
1974	1786
1975	1885
1976	2085 (includes 189 Hawk XP)
1977	2309 (includes 598 Hawk XP)
1978	2023 (includes 213 Hawk XP)
1979	2120 (includes 254 Hawk XP, 185 Cutlass RG)
1980	1362 (includes 155 Hawk XP, 185 Cutlass RG)
1981	1250 (includes 40 Hawk XP, 271 Cutlass RG)
1982	437 (includes 3 Hawk XP, 107 Cutlass RG, 8 Cutlass)
1983	305 (includes 54 Cutlass RG, 20 Cutlass)
1984	235 (includes 40 Cutlass RG, 8 Cutlass)
1985	209 (includes 15 Cutlass RG)
1986	116 (includes 1 Cutlass RG)
1987	no production

Model 175

Year

1958	702
1959	727
1960	501
1961	126
1962	50
1963	13

SERIAL NUMBERS

The following list gives the serial number ranges for all years and models of the 172 airplanes. Starting in 1961, the numeral "172" was added to the front of all Model 172 serial numbers (17271035); Cutlass Q Models share the standard 172 serial number sequence. Note that for certain years, apparently more serial numbers were issued than airplanes built.

Model 172

Year	Beginning	Ending
1956	28000	29174
1957	29175	29999
1957	36000	36215
1958	36216	36965
1959	36966	36999
1959	46001	46754
1960	46755	47746
1961	47747	48734
1962	48735	49544
1963	49545	50572
1964	50573	51822
1965	51823	53392
1966	53393	54892
1967	54893	56512
1968	56513	57161
1969	57162	58486
1970	58487	59223
1971	59224	59903
1972	59904	60758
1973	60759	61898
1974	61899	63458
1975	63459	65684
1976	65685	67584
1977	67585	69309
1978	69310	71034
1979	71035	72884
1980	72885	74009
1981	74010	75034
1982	75035	75759
1983	75760	76079
1984	76080	76259
1985	76260	76516
1986	76517	76673

Model 175

Year	Beginning	Ending
1958	55001	55703
1959	55704	56238
1960	56239	56777
1961	56778	57002
1962	57003	57119

Powermatic

Year	Beginning	Ending
1963	P17257120	P17257189

Hawk XP

Year	Beginning	Ending
1977	R1722000	R1722724 (some built in 1976)
1978	R1722725	R1722929
1979	R1722930	R1723199
1980	R1723200	R1723399
1981	R1723400	R1723454 (some sold as 1982s)

Cutlass RG

Year	Beginning	Ending
1980	RG0001	RG0570 (some built in 1979)
1981	RG0571	RG0890
1982	RG0891	RG1099
1983	RG1100	RG1144
1984	RG1145	RG1177
1985	RG1178	RG1191 (some sold as 1986s)

Cutlass Q

Year	Beginning	Ending
1983	75869	76079 (some built in 1982)
1984	76080	76259

TIME FLIES

Due to the high prices and general lack of production of Cessna and other similar size airplanes, the FAA estimates that the average four-place single-engine airplane is in excess of 25 years old. Does this sound possible? Remember that 1955, when the 172 was first available, was almost 40 years ago. Time flies.

2

Specifications

NUMEROUS VERSIONS OF THE 172 are reviewed in this chapter. Notice the differences between the various year models and compare the differences to the model changes described in chapter 1 (Fig. 2-1 through Fig. 2-6). Specifications are based upon material from Cessna Aircraft Company.

Fig. 2-1. The early 172s have straight tails and are referred to as "straight tails" or "straight-tailed fastbacks." "Fastback" refers to the lack of rear cabin windows.

1956

Model: 172

Speed
    Top Speed at Sea Level: 135 mph
    Cruise: 124 mph
Range at Recommended Cruise
    519 mi

Fig. 2-2. This 1962 172 is a "swept-tail fastback."

Fig. 2-3. Built in 1964, this 172 is typical of most 172s and Skyhawks seen today.

    4.2 hrs
    124 mph
Maximum Range at 7500 ft
    620 mi
    6.4 hrs
    97 mph
Performance
    Rate of Climb at Sea Level: 660 fpm
    Service Ceiling: 13,300 ft

Fig. 2-4. Notice the unfaired landing gear. This is a 172, not a Skyhawk, and therefore somewhat plainer in appearance.

Fig. 2-5. In 1977, the "100" versions of the Skyhawk appeared, indicating 100LL avgas burned by the troublesome O-320-H2AD engine.

Capacities
    Standard Fuel: 42 gal
Engine
    Continental O-300A
    TBO: 1500 hrs
    Power: 145 hp
Dimensions
    Wingspan: 36 ft 00 in
    Wing Area: (sq ft) 175

Fig. 2-6. Compare this late-model Skyhawk with earlier versions. Notice the one-piece wind-shield, the lower stance of the landing gear, and a stylish paint scheme.

Wing Loading: (lbs/sq ft) 12.6
Power Loading: (lbs/hp) 15.2
Length: 25 ft 00 in
Height: 8 ft 06 in
Weight
    Gross: 2200 lbs
    Empty: 1260 lbs
    Useful Load: 940 lbs
    Baggage: 120 lbs

1957

Model: 172

Speed
    Top Speed at Sea Level: 135 mph
    Cruise: 124 mph
Range at Recommended Cruise
    519 mi
    4.2 hrs
    124 mph
Maximum Range at 7500 ft
    620 mi
    6.4 hrs
    97 mph
Performance
    Rate of Climb at Sea Level: 660 fpm
    Service Ceiling: 13,300 ft
Capacities
    Standard Fuel: 42 gal

Engine
Continental O-300A
TBO: 1500 hrs
Power: 145 hp
Dimensions
Wingspan: 36 ft 00 in
Wing Area: (sq ft) 175
Wing Loading: (lbs/sq ft) 12.6
Power Loading: (lbs/hp) 15.2
Length: 25 ft 00 in
Height: 8 ft 06 in
Weights
Gross: 2200 lbs
Empty: 1260 lbs
Useful Load: 940 lbs
Baggage: 120 lbs

1958

Model: 172

Speed
Top Speed at Sea Level: 135 mph
Cruise: 124 mph
Range at Recommended Cruise
519 mi
4.2 hrs
124 mph
Maximum Range at 7500 ft
620 mi
6.4 hrs
97 mph
Performance
Rate of Climb at Sea Level: 660 fpm
Service Ceiling: 13,300 ft
Capacities
Standard Fuel: 42 gal
Engine
Continental O-300A
TBO: 1500 hrs
Power: 145 hp
Dimensions
Wingspan: 36 ft 00 in
Wing Area: (sq ft) 175
Wing Loading: (lbs/sq ft) 12.6
Power Loading: (lbs/hp) 15.2
Length: 25 ft 00 in
Height: 8 ft 06 in

Weights
    Gross: 2200 lbs
    Empty: 1260 lbs
    Useful Load: 940 lbs
    Baggage: 120 lbs

1959
Model: 172

Speed
    Top Speed at Sea Level: 135 mph
    Cruise: 124 mph
Range at Recommended Cruise
    519 mi
    4.2 hrs
    124 mph
Maximum Range at 7500 ft
    620 mi
    6.4 hrs
    97 mph
Performance
    Rate of Climb at Sea Level: 660 fpm
    Service Ceiling: 13,300 ft
Capacities
    Standard Fuel: 42 gal
Engine
    Continental O-300A
    TBO: 1500 hrs
    Power: 145 hp
Dimensions
    Wingspan: 36 ft 00 in
    Wing Area: (sq ft) 175
    Wing Loading: (lbs/sq ft) 12.6
    Power Loading: (lbs/hp) 15.2
    Length: 25 ft 00 in
    Height: 8 ft 06 in
Weights
    Gross: 2200 lbs
    Empty: 1260 lbs
    Useful Load: 940 lbs
    Baggage: 120 lbs

1960
Model: 172A

Speed
    Top Speed at Sea Level: 140 mph

Cruise, 70 percent power at 8000 ft: 131 mph
Range
    Cruise, 70 percent power at 8000 ft: 545 mi
        with 37 Gallons, No Reserve: 4.2 hrs at 131 mph
    Maximum Range at 10,000 ft: 790 mi
        with 37 Gallons, No Reserve: 8.3 hrs at 95 mph
Performance
    Rate of Climb at Sea Level: 730 fpm
    Service Ceiling: 15,100 ft
    Takeoff
        Ground Run: 780 ft
        Over 50-ft Obstacle: 1370 ft
    Landing
        Ground Roll: 680 ft
        Over 50-ft Obstacle: 1115 ft
Capacities
    Standard Fuel: 42 gal
    Oil: 8 qts
Engine
    Continental O-300C
    TBO: 1800 hrs
    Power: 145 hp
    Propeller: 1C171/EM 7654 McCauley
Dimensions
    Wingspan: 36 ft 00 in
    Wing Area: (sq ft) 174
    Wing Loading: (lbs/sq ft) 12.6
    Power Loading: (lbs/hp) 15.2
    Length: 26 ft 04 in
    Height: 8 ft 11 in
Weights
    Gross: 2200 lbs
    Empty: 1252 lbs
    Useful Load: 948 lbs
    Baggage: 120 lbs

1961

Model: 172B

Speed
    Top Speed at Sea Level: 139 mph
    Cruise, 75 percent power at 7000 ft: 131 mph
Range
    Cruise, 75 percent power at 7000 ft: 535 mi
        39 Gallons, No Reserve: 4.1 hrs at 131 mph
    Optimum Range at 10,000 ft: 780 mi
        39 Gallons, No Reserve: 7.8 hrs at 100 mph

Performance
    Rate of Climb at Sea Level: 700 fpm
    Service Ceiling: 14,550 ft
    Takeoff
        Ground Run: 825 ft
        Over 50-ft Obstacle: 1430 ft
    Landing
        Ground Roll: 690 ft
        Over 50-ft Obstacle: 1140 ft
Capacities
    Standard Fuel: 42 gal
    Oil Capacities: 8 qts
Engine
    Continental O-300C
    TBO: 1800 hrs
    Power: 145 hp
    Propeller: (diameter) 76 in
Dimensions
    Wingspan: 36 ft 02 in
    Wing Area: (sq ft) 175
    Wing Loading: (lbs/sq ft) 12.9
    Power Loading: (lbs/hp) 15.5
    Length: 26 ft 06 in
    Height: 8 ft 11 in
Weights
    Gross: 2250 lbs
    Empty: 1260 lbs
    Useful Load: 990 lbs
    Baggage: 120 lbs

Model: 172B Skyhawk

Speed
    Top Speed at Sea Level: 140 mph
    Cruise, 75 percent power at 7000 ft: 131 mph
Range
    Cruise, 75 percent power at 7000 ft: 540 mi
        39 Gallons, No Reserve: 4.1 hrs at 131 mph
    Optimum Range at 10,000 ft: 780 mi
        39 Gallons, No Reserve: 7.8 hrs at 100 mph
Performance
    Rate of Climb at Sea Level: 730 fpm
    Service Ceiling: 15,100 ft
    Takeoff
        Ground Run: 875 ft
        Over 50-ft Obstacle: 1370 ft
    Landing
        Ground Roll: 600 ft

Over 50-ft Obstacle: 1115 ft
Capacities
    Standard Fuel: 42 gal
    Oil: 8 qts
Engine
    Continental O-300D
    TBO: 1800 hrs
    Power: 145 hp
    Propeller: (diameter) 76 in
Dimensions
    Wingspan: 36 ft 02 in
    Wing Area: (sq ft) 175
    Wing Loading: (lbs/sq ft) 12.9
    Power Loading: (lbs/hp) 15.5
    Length: 26 ft 06 in
    Height: 8 ft 11 in
Weights
    Gross: 2200 lbs
    Empty: 1325 lbs
    Useful Load: 875 lbs
    Baggage: 120 lbs

1962

Model: 172C

Speed
    Top Speed at Sea Level: 139 mph
    Cruise, 75 percent power at 7000 ft: 131 mph
Range
    Cruise, 75 percent power at 7000 ft: 535 mi
        39 Gallons, No Reserve: 4.1 hrs at 131 mph
    Optimum Range at 10,000 ft: 780 mi
        39 Gallons, No Reserve: 7.8 hrs at 100 mph
Performance
    Rate of Climb at Sea Level: 700 fpm
    Service Ceiling: 14,550 ft
    Takeoff
        Ground Run: 825 ft
        Over 50-ft Obstacle: 1430 ft
    Landing
        Ground Roll: 690 ft
        Over 50-ft Obstacle: 1140 ft
Capacities
    Standard Fuel: 42 gal
    Oil: 8 qts
Engine
    Continental O-300C

TBO: 1800 hrs
Power: 145 hp
Propeller: (diameter) 76 in
Dimensions
Wingspan: 36 ft 02 in
Wing Area: (sq ft) 175
Wing Loading: (lbs/sq ft) 12.9
Power Loading: (lbs/hp) 15.5
Length: 26 ft 06 in
Height: 8 ft 11 in
Weights
Gross: 2250 lbs
Empty: 1260 lbs
Useful Load: 990 lbs
Baggage: 120 lbs

Model: 172C Skyhawk

Speed
Top Speed at Sea Level: 140 mph
Cruise, 75 percent power at 7000 ft: 132 mph
Range
Cruise, 75 percent power at 7000 ft: 540 mi
39 Gallons, No Reserve: 4.1 hrs at 132 mph
Optimum Range at 10,000 ft: 780 mi
39 Gallons, No Reserve: 7.8 hrs at 100 mph
Performance
Rate of Climb at Sea Level: 700 fpm
Service Ceiling: 14,550 ft
Takeoff
Ground Run: 825 ft
Over 50-ft Obstacle: 1430 ft
Landing
Ground Roll: 690 ft
Over 50-ft Obstacle: 1140 ft
Capacities
Standard Fuel: 42 gal
Oil: 8 qts
Engine
Continental O-300D
TBO: 1800 hrs
Power: 145 hp
Propeller: (diameter) 76 in
Dimensions
Wingspan: 36 ft 02 in
Wing Area: (sq ft) 175
Wing Loading: (lbs/sq ft) 12.9
Power Loading: (lbs/hp) 15.5

Dimensions
    Length: 26 ft 06 in
    Height: 8 ft 11 in
Weights
    Gross: 2250 lbs
    Empty: 1330 lbs
    Useful Load: 990 lbs
    Baggage: 120 lbs

1963

Model: 172D

Speed
    Top Speed at Sea Level: 138 mph
    Cruise, 75 percent power at 7000 ft: 130 mph
Range
    Cruise, 75 percent power at 7000 ft: 595 mi
        39 Gallons, No Reserve: 4.6 hrs at 130 mph
    Optimum Range at 10,000 ft: 720 mi
        39 Gallons, No Reserve: 7.1 hrs at 102 mph
Performance
    Rate of Climb at Sea Level: 645 fpm
    Service Ceiling: 13,100 ft
    Takeoff
        Ground Run: 865 ft
        Over 50-ft Obstacle: 1525 ft
    Landing
        Ground Roll: 520 ft
        Over 50-ft Obstacle: 1250 ft
Capacities
    Standard Fuel: 42 gal
    Oil: 8 qts
Engine
    Continental O-300C
    TBO: 1800 hrs
    Power: 145 hp
    Propeller: (diameter) 76 in
Dimensions
    Wingspan: 36 ft 02 in
    Wing Area: (sq ft) 174
    Wing Loading: (lbs/sq ft) 13.2
    Power Loading: (lbs/hp) 15.9
    Length: 26 ft 06 in
    Height: 8 ft 11 in
Weights
    Gross: 2300 lbs
    Empty: 1260 lbs

Useful Load: 1040 lbs
Baggage: 120 lbs

Model: 172D Skyhawk

Speed
  Top Speed at Sea Level: 139 mph
  Cruise, 75 percent power at 7000 ft: 131 mph
Range
  Cruise, 75 percent power at 7000 ft: 600 mi
        39 Gallons, No Reserve: 4.6 hrs at 131 mph
  Optimum Range at 10,000 ft: 720 mi
        39 Gallons, No Reserve: 7.1 hrs at 102 mph
Performance
  Rate of Climb at Sea Level: 645 fpm
  Service Ceiling: 13,100 ft
  Takeoff
      Ground Run: 865 ft
      Over 50-ft Obstacle: 1525 ft
  Landing
      Ground Roll: 520 ft
      Over 50-ft Obstacle: 1250 ft
Capacities
  Standard Fuel: 42 gal
  Oil: 8 qts
Engine
  Continental O-300D
  TBO: 1800 hrs
  Power: 145 hp
  Propeller: (diameter) 76 in
Dimensions
  Wingspan: 36 ft 02 in
  Wing Area: (sq ft) 174
  Wing Loading: (lbs/sq ft) 13.2
  Power Loading: (lbs/hp) 15.9
  Length: 26 ft 06 in
  Height: 8 ft 11 in
Weights
  Gross: 2300 lbs
  Empty: 1330 lbs
  Useful Load: 970 lbs
  Baggage: 120 lbs

Model: 172 Powermatic

Speed
  Top Speed at Sea Level: 150 mph
  Cruise, 75 percent power at 7000 ft: 142 mph

Range
   Cruise, 75 percent power at 7000 ft: 550 mi
          41.5 Gallons, No Reserve: 3.9 hrs at 142 mph
   Optimum Range at 10,000 ft: 720 mi
          41.5 Gallons, No Reserve: 6.9 hrs at 104 mph
Performance
   Rate of Climb at Sea Level: 950 fpm
   Service Ceiling: 17,800 ft
   Takeoff
      Ground Run: 600 ft
      Over 50-ft Obstacle: 1205 ft
   Landing
      Ground Roll: 610 ft
      Over 50-ft Obstacle: 1200 ft
Capacities
   Standard Fuel: 52 gal
   Oil: 10 qts
Engine
   Continental GO-300E
   TBO: 1200 hrs
   Power: 175 hp
   Propeller: (diameter)C/S 84 in
Dimensions
   Wingspan: 36 ft 02 in
   Wing Area: (sq ft) 175
   Wing Loading: (lbs/sq ft) 14.1
   Power Loading: (lbs/hp) 14.0
   Length: 26 ft 06 in
   Height: 8 ft 11 in
Weights
   Gross: 2450 lbs
   Empty: 1410 lbs
   Useful Load: 1040 lbs
   Baggage: 120 lbs

1964

Model: 172E

Speed
   Top Speed at Sea Level: 138 mph
   Cruise, 75 percent power at 7000 ft: 130 mph
Range
   Cruise, 75 percent power at 7000 ft: 595 mi
          39 Gallons, No Reserve: 4.6 hrs at 130 mph
   Optimum Range at 10,000 ft: 720 mi
          39 Gallons, No Reserve: 7.1 hrs at 102 mph

Performance
    Rate of Climb at Sea Level: 645 fpm
    Service Ceiling: 13,100 ft
    Takeoff
        Ground Run: 865 ft
        Over 50-ft Obstacle: 1525 ft
    Landing
        Ground Roll: 520 ft
        Over 50-ft Obstacle: 1250 ft
Capacities
    Standard Fuel: 42 gal
    Oil: 8 qts
Engine
    Continental O-300C
    TBO: 1800 hrs
    Power: 145 hp
    Propeller: (diameter) 76 in
Dimensions
    Wingspan: 36 ft 02 in
    Wing Area: (sq ft) 174
    Wing Loading: (lbs/sq ft) 13.2
    Power Loading: (lbs/hp) 15.9
    Length: 26 ft 06 in
    Height: 8 ft 11 in
Weights
    Gross: 2300 lbs
    Empty: 1255 lbs
    Useful Load: 1045 lbs
    Baggage: 120 lbs

Model: 172E Skyhawk

Speed
    Top Speed at Sea Level: 139 mph
    Cruise, 75 percent power at 7000 ft: 131 mph
Range
    Cruise, 75 percent power at 7000 ft: 600 mi
            39 Gallons, No Reserve: 4.6 hrs at 131 mph
    Optimum Range at 10,000 ft: 720 mi
            39 Gallons, No Reserve: 7.1 hrs at 102 mph
Performance
    Rate of Climb at Sea Level: 645 fpm
    Service Ceiling: 13,100 ft
    Takeoff
        Ground Run: 865 ft
        Over 50-ft Obstacle: 1525 ft
    Landing
        Ground Roll: 520 ft

Over 50-ft Obstacle: 1250 ft

Capacities
  Standard Fuel: 42 gal
  Oil: 8 qts
Engine
  Continental O-300D
  TBO: 1800 hrs
  Power: 145 hp
  Propeller: (diameter) 76 in
Dimensions
  Wingspan: 36 ft 02 in
  Wing Area: (sq ft) 174
  Wing Loading: (lbs/sq ft) 13.2
  Power Loading: (lbs/hp) 15.9
  Length: 26 ft 06 in
  Height: 8 ft 11 in
Weights
  Gross: 2300 lbs
  Empty: 1320 lbs
  Useful Load: 980 lbs
  Baggage: 120 lbs

1965

Model: 172F

Speed
  Top Speed at Sea Level: 138 mph
  Cruise, 75 percent power at 7000 ft: 130 mph
Range
  Cruise, 75 percent power at 7000 ft: 595 mi
        39 Gallons, No Reserve: 4.6 hrs at 130 mph
  Optimum Range at 10,000 ft: 720 mi
        39 Gallons, No Reserve: 7.1 hrs at 102 mph
Performance
  Rate of Climb at Sea Level: 645 fpm
  Service Ceiling: 13,100 ft
  Takeoff
    Ground Run: 865 ft
    Over 50-ft Obstacle: 1525 ft
  Landing
    Ground Roll: 520 ft
    Over 50-ft Obstacle: 1250 ft
Capacities
  Standard Fuel: 42 gal
  Oil: 8 qts
Engine
  Continental O-300C

TBO: 1800 hrs
Power: 145 hp
Propeller: (diameter) 76 in
Dimensions
Wingspan: 36 ft 02 in
Wing Area: (sq ft) 174
Wing Loading: (lbs/sq ft) 13.2
Power Loading: (lbs/hp) 15.9
Length: 26 ft 06 in
Height: 8 ft 11 in
Weights
Gross: 2300 lbs
Empty: 1260 lbs
Useful Load: 1040 lbs
Baggage: 120 lbs

Model: 172F Skyhawk

Speed
Top Speed at Sea Level: 139 mph
Cruise, 75 percent power at 7000 ft: 131 mph
Range
Cruise, 75 percent power at 7000 ft: 600 mi
39 Gallons, No Reserve: 4.6 hrs at 131 mph
Optimum Range at 10,000 ft: 720 mi
39 Gallons, No Reserve: 7.1 hrs at 102 mph
Performance
Rate of Climb at Sea Level: 645 fpm
Service Ceiling: 13,100 ft
Takeoff
Ground Run: 865 ft
Over 50-ft Obstacle: 1525 ft
Landing
Ground Roll: 520 ft
Over 50-ft Obstacle: 1250 ft
Capacities
Standard Fuel: 42 gal
Oil: 8 qts
Engine
Continental O-300D
TBO: 1800 hrs
Power: 145 hp
Propeller: (diameter) 76 in
Dimensions
Wingspan: 36 ft 02 in
Wing Area: (sq ft) 174
Wing Loading: (lbs/sq ft) 13.2
Power Loading: (lbs/hp) 15.9

Length: 26 ft 06 in
Height: 8 ft 11 in
Weights
 Gross: 2300 lbs
 Empty: 1320 lbs
 Useful Load: 980 lbs
 Baggage: 120 lbs

1966

Model: 172G

Speed
 Top Speed at Sea Level: 138 mph
 Cruise, 75 percent power at 7000 ft: 130 mph
Range
 Cruise, 75 percent power at 7000 ft: 595 mi
  39 Gallons, No Reserve: 4.6 hrs at 130 mph
 Optimum Range at 10,000 ft: 720 mi
  39 Gallons, No Reserve: 7.1 hrs at 102 mph
Performance
 Rate of Climb at Sea Level: 645 fpm
 Service Ceiling: 13,100 ft
 Takeoff
  Ground Run: 865 ft
  Over 50-ft Obstacle: 1525 ft
 Landing
  Ground Roll: 520 ft
  Over 50-ft Obstacle: 1250 ft
Capacities
 Standard Fuel: 42 gal
 Oil: 8 qts
Engine
 Continental O-300C
 TBO: 1800 hrs
 Power: 145 hp
 Propeller: (diameter) 76 in
Dimensions
 Wingspan: 36 ft 02 in
 Wing Area: (sq ft) 174
 Wing Loading: (lbs/sq ft) 13.2
 Power Loading: (lbs/hp) 15.9
 Length: 26 ft 06 in
 Height: 8 ft 11 in
Weights
 Gross: 2300 lbs
 Empty: 1260 lbs
 Useful Load: 1040 lbs

Baggage: 120 lbs

Model: 172G Skyhawk

Speed
  Top Speed at Sea Level: 139 mph
  Cruise, 75 percent power at 7000 ft: 131 mph
Range
  Cruise, 75 percent power at 7000 ft: 600 mi
        39 Gallons, No Reserve: 4.6 hrs at 131 mph
  Optimum Range at 10,000 ft: 720 mi
        39 Gallons, No Reserve: 7.1 hrs at 102 mph
Performance
  Rate of Climb at Sea Level: 645 fpm
  Service Ceiling: 13,100 ft
  Takeoff
      Ground Run: 865 ft
      Over 50-ft Obstacle: 1525 ft
  Landing
      Ground Roll: 520 ft
      Over 50-ft Obstacle: 1250 ft
Capacities
  Standard Fuel: 42 gal
  Oil: 8 qts
Engine
  Continental O-300D
  TBO: 1800 hrs
  Power: 145 hp
  Propeller: (diameter) 76 in
Dimensions
  Wingspan: 36 ft 02 in
  Wing Area: (sq ft) 174
  Wing Loading: (lbs/sq ft) 13.2
  Power Loading: (lbs/hp) 15.9
  Length: 26 ft 06 in
  Height: 8 ft 11 in
Weights
  Gross: 2300 lbs
  Empty: 1320 lbs
  Useful Load: 980 lbs
  Baggage: 120 lbs

1967

Model: 172H

Speed
  Top Speed at Sea Level: 138 mph

Cruise, 75 percent power at 7000 ft: 130 mph
Range
    Cruise, 75 percent power at 7000 ft: 595 mi
          39 Gallons, No Reserve: 4.6 hrs at 130 mph
    Optimum Range at 10,000 ft: 720 mi
          39 Gallons, No Reserve: 7.1 hrs at 102 mph
Performance
    Rate of Climb at Sea Level: 645 fpm
    Service Ceiling: 13,100 ft
    Takeoff
        Ground Run: 865 ft
        Over 50-ft Obstacle: 1525 ft
    Landing
        Ground Roll: 520 ft
        Over 50-ft Obstacle: 1250 ft
Capacities
    Standard Fuel: 42 gal
    Oil: 8 qts
Engine
    Continental O-300D
    TBO: 1800 hrs
    Power: 145 hp
    Propeller: (diameter) 76 in
Dimensions
    Wingspan: 36 ft 02 in
    Wing Area: (sq ft) 174
    Wing Loading: (lbs/sq ft) 13.2
    Power Loading: (lbs/hp) 15.87
    Length: 26 ft 11 in
    Height: 8 ft 09 in
Weights
    Gross: 2300 lbs
    Empty: 1275 lbs
    Useful Load: 1025 lbs
    Baggage: 120 lbs

Model: 172H Skyhawk

Speed
    Top Speed at Sea Level: 139 mph
    Cruise, 75 percent power at 7000 ft: 131 mph
Range
    Cruise, 75 percent power at 7000 ft: 600 mi
          39 Gallons, No Reserve: 4.6 hrs at 131 mph
    Optimum Range at 10,000 ft: 720 mi
          39 Gallons, No Reserve: 7.1 hrs at 102 mph
Performance
    Rate of Climb at Sea Level: 645 fpm

Service Ceiling: 13,100 ft
Takeoff
Ground Run: 865 ft
Over 50-ft Obstacle: 1525 ft
Landing
Ground Roll: 520 ft
Over 50-ft Obstacle: 1250 ft
Capacities
Standard Fuel: 42 gal
Oil: 8 qts
Engine
Continental O-300D
TBO: 1800 hrs
Power: 145 hp
Propeller: (diameter) 76 in
Dimensions
Wingspan: 36 ft 02 in
Wing Area: (sq ft) 174
Wing Loading: (lbs/sq ft) 13.2
Power Loading: (lbs/hp) 15.87
Length: 26 ft 11 in
Height: 8 ft 09 in
Weight
Gross: 2300 lbs
Empty: 1340 lbs
Useful Load: 960 lbs
Baggage: 120 lbs

1968

Model: 172I

Speed
Top Speed at Sea Level: 139 mph
Cruise, 75 percent power at 9000 ft: 131 mph
Range
Cruise, 75 percent power at 9000 ft: 615 mi
38 Gallons, No Reserve: 4.7 hrs at 131 mph
Optimum Range at 10,000 ft: 640 mi
5.5 hrs at 117 mph
Performance
Rate of Climb at Sea Level: 645 fpm
Service Ceiling: 13,100 ft
Takeoff
Ground Run: 865 ft
Over 50-ft Obstacle: 1525 ft
Landing
Ground Roll: 520 ft

Over 50-ft Obstacle: 1250 ft
Capacities
    Standard Fuel: 42 gal
    Oil: 8 qts
Engine
    Lycoming O-320-E2D
    TBO: 2000 hrs
    Power: 150 hp
Dimensions
    Wing Loading: (lbs/sq ft) 13.2
    Power Loading: (lbs/hp) 15.33
Weights
    Gross: 2300 lbs
    Empty: 1230 lbs
    Useful Load: 1070 lbs
    Baggage: 120 lbs

Model: 172I Skyhawk

Speed
    Top Speed at Sea Level: 140 mph
    Cruise, 75 percent power at 9000 ft: 132 mph
Range
    Cruise, 75 percent power at 9000 ft: 620 mi
            38 Gallons, No Reserve: 4.7 hrs at 132 mph
Optimum Range at 10,000 ft: 655 mi
            5.5 hrs at 118 mph
Performance
    Rate of Climb at Sea Level: 645 fpm
    Service Ceiling: 13,100 ft
    Takeoff
        Ground Run: 865 ft
        Over 50-ft Obstacle: 1525 ft
    Landing
        Ground Roll: 520 ft
        Over 50-ft Obstacle: 1250 ft
Capacities
    Standard Fuel: 42 gal
    Oil: 8 qts
Engine
    Lycoming O-320-E2D
    TBO: 2000 hrs
    Power: 150 hp
Dimensions
    Wing Loading: (lbs/sq ft) 13.2
    Power Loading: (lbs/hp) 15.33
Weights
    Gross: 2300 lbs

Empty: 1300 lbs
Useful Load: 1000 lbs
Baggage: 120 lbs

1969

Model: 172K

Speed
    Top Speed at Sea Level: 139 mph
    Cruise, 75 percent power at 9000 ft: 131 mph
Range
    Cruise, 75 percent power at 9000 ft: 615 mi
            38 Gallons, No Reserve: 4.7 hrs at 131 mph
    Optimum Range at 10,000 ft: 640 mi
            5.5 hrs at 117 mph
Performance
    Rate of Climb at Sea Level: 645 fpm
    Service Ceiling: 13,100 ft
    Takeoff
        Ground Run: 865 ft
        Over 50-ft Obstacle: 1525 ft
    Landing
        Ground Roll: 520 ft
        Over 50-ft Obstacle: 1250 ft
Capacities
    Standard Fuel: 42 gal
    Oil: 8 qts
Engine
    Lycoming O-320-E2D
    TBO: 2000 hrs
    Power: 150 hp
Dimensions
    Wing Loading: (lbs/sq ft) 13.2
    Power Loading: (lbs/hp) 15.33
Weights
    Gross: 2300 lbs
    Empty: 1230 lbs
    Useful Load: 1070 lbs
    Baggage: 120 lbs

Model: 172K Skyhawk

Speed
    Top Speed at Sea Level: 140 mph
    Cruise, 75 percent power at 9000 ft: 132 mph
Range
    Cruise, 75 percent power at 9000 ft: 620 mi

38 Gallons, No Reserve: 4.7 hrs at 132 mph
Optimum Range at 10,000 ft: 655 mi
5.5 hrs at 118 mph

Performance
    Rate of Climb at Sea Level: 645 fpm
    Service Ceiling: 13,100 ft
    Takeoff
        Ground Run: 865 ft
        Over 50-ft Obstacle: 1525 ft
    Landing
        Ground Roll: 520 ft
        Over 50-ft Obstacle: 1250 ft

Capacities
    Standard Fuel: 42 gal
    Oil: 8 qts

Engine
    Lycoming O-320-E2D
    TBO: 2000 hrs
    Power: 150 hp

Dimensions
    Wing Loading: (lbs/sq ft) 13.2
    Power Loading: (lbs/hp) 15.33

Weights
    Gross: 2300 lbs
    Empty: 1300 lbs
    Useful Load: 1000 lbs
    Baggage: 120 lbs

1970

Model: 172K

Speed
    Top Speed at Sea Level: 139 mph
    Cruise, 75 percent power at 9000 ft: 131 mph

Range
    Cruise, 75 percent power at 9000 ft: 615 mi
        38 Gallons, No Reserve: 4.7 hrs at 131 mph
    Cruise, 75 percent at 9000 ft: 775 mi
        48 Gallons, No Reserve: 5.9 hrs at 131 mph
    Optimum Range at 10,000 ft: 640 mi
        38 Gallons, No Reserve: 5.5 hrs at 117 mph
    Optimum Range at 10,000 ft: 820 mi
        48 Gallons, No Reserve: 7.0 hrs at 117 mph

Performance
    Rate of Climb at Sea Level: 645 fpm
    Service Ceiling: 13,100 ft
    Takeoff

Ground Run: 865 ft
Over 50-ft Obstacle: 1525 ft
Landing
Ground Roll: 520 ft
Over 50-ft Obstacle: 1250 ft
Stall Speed
Flaps Up, Power Off: 57 mph
Flaps Down, Power Off: 49 mph
Capacities
Standard Fuel: 42 gal
w/optional tanks: 52 gal
Oil: 8 qts
Engine
Lycoming O-320-E2D
TBO: 2000 hrs
Power: (at 2700 rpm) 150 hp
Propeller: (diameter) 76 in
Dimensions
Wingspan: 35 ft 09 in
Wing Area: (sq ft) 174
Wing Loading: (lbs/sq ft) 13.2
Power Loading: (lbs/hp) 15.3
Length: 26 ft 11 in
Height: 8 ft 09 in
Weights
Gross: 2300 lbs
Empty: 1245 lbs
Useful Load: 1055 lbs
Baggage: 120 lbs

Model: 172K Skyhawk

Speed
Top Speed at Sea Level: 140 mph
Cruise, 75 percent power at 9000 ft: 132 mph
Range
Cruise, 75 percent power at 9000 ft: 620 mi
38 Gallons, No Reserve: 4.7 hrs at 132 mph
Cruise, 75 percent power at 9000 ft: 780 mi
48 Gallons, No Reserve: 5.9 hrs at 132 mph
Optimum Range at 10,000 ft: 655 mi
38 Gallons, No Reserve: 5.5 hrs at 118 mph
Optimum Range at 10,000 ft: 830 mi
48 Gallons, No Reserve: 7.0 hrs at 118 mph
Performance
Rate of Climb at Sea Level: 645 fpm
Service Ceiling: 13,100 ft
Takeoff

Ground Run: 865 ft
Over 50-ft Obstacle: 1525 ft
Landing
Ground Roll: 520 ft
Over 50-ft Obstacle: 1250 ft
Stall Speed
Flaps Up, Power Off: 57 mph
Flaps Down, Power Off: 49 mph
Capacities
Standard Fuel: 42 gal
w/optional tanks: 52 gal
Oil: 8 qts
Engine
Lycoming O-320-E2D
TBO: 2000 hrs
Power: 150 hp
Propeller: (diameter) 76 in
Dimensions
Wingspan: 35 ft 09 in
Wing Area: (sq ft) 174
Wing Loading: (lbs/sq ft) 13.2
Power Loading: (lbs/hp) 15.3
Length: 26 ft 11 in
Height: 8 ft 09 in
Weights
Gross: 2300 lbs
Empty: 1315 lbs
Useful Load: 985 lbs
Baggage: 120 lbs

1971

Model: 172L

Speed
Top Speed at Sea Level: 139 mph
Cruise, 75 percent power at 9000 ft: 131 mph
Range
Cruise, 75 percent power at 9000 ft: 615 mi
38 Gallons, No Reserve: 4.7 hrs at 131 mph
Cruise, 75 percent at 9000 ft: 775 mi
48 Gallons, No Reserve: 5.9 hrs at 131 mph
Optimum Range at 10,000 ft: 640 mi
38 Gallons, No Reserve: 5.5 hrs at 117 mph
Optimum Range at 10,000 ft: 820 mi
48 Gallons, No Reserve: 7.0 hrs at 117 mph
Performance
Rate of Climb at Sea Level: 645 fpm

Service Ceiling: 13,100 ft
Takeoff
    Ground Run: 865 ft
    Over 50-ft Obstacle: 1525 ft
Landing
    Ground Roll: 520 ft
    Over 50-ft Obstacle: 1250 ft
Stall Speed
    Flaps Up, Power Off: 57 mph
    Flaps Down, Power Off: 49 mph
Capacities
    Standard Fuel: 42 gal
    w/optional tanks: 52 gal
    Oil: 8 qts
Engine
    Lycoming O-320-E2D
    TBO: 2000 hrs
    Power: (at 2700 rpm) 150 hp
    Propeller: (diameter) 76 in
Dimensions
    Wingspan: 35 ft 09 in
    Wing Area: (sq ft) 174
    Wing Loading: (lbs/sq ft) 13.2
    Power Loading: (lbs/hp) 15.3
    Length: 26 ft 11 in
    Height: 8 ft 09 in
Weights
    Gross: 2300 lbs
    Empty: 1250 lbs
    Useful Load: 1050 lbs
    Baggage: 120 lbs

Model: 172L Skyhawk

Speed
    Top Speed at Sea Level: 140 mph
    Cruise, 75 percent power at 9000 ft: 132 mph
Range
    Cruise, 75 percent power at 9000 ft: 620 mi
        38 Gallons, No Reserve: 4.7 hrs at 132 mph
    Cruise, 75 percent power at 9000 ft: 780 mi
        48 Gallons, No Reserve: 5.9 hrs at 132 mph
    Optimum Range at 10,000 ft: 655 mi
        38 Gallons, No Reserve: 5.5 hrs at 118 mph
    Optimum Range at 10,000 ft: 830 mi
        48 Gallons, No Reserve: 7.0 hrs at 118 mph
Performance
    Rate of Climb at Sea Level: 645 fpm

Service Ceiling: 13,100 ft
Takeoff
    Ground Run: 865 ft
    Over 50-ft Obstacle: 1525 ft
Landing
    Ground Roll: 520 ft
    Over 50-ft Obstacle: 1250 ft
Stall Speed
    Flaps Up, Power Off: 57 mph
    Flaps Down, Power Off: 49 mph
Capacities
    Standard Fuel: 42 gal
    w/optional tanks: 52 gal
    Oil: 8 qts
Engine
    Lycoming O-320-E2D
    TBO: 2000 hrs
    Power: (at 2700 rpm) 150 hp
    Propeller: (diameter) 76 in
Dimensions
    Wingspan: 35 ft 09 in
    Wing Area: (sq ft) 174
    Wing Loading: (lbs/sq ft) 13.2
    Power Loading: (lbs/hp) 15.3
    Length: 26 ft 11 in
    Height: 8 ft 09 in
Weights
    Gross: 2300 lbs
    Empty: 1300 lbs
    Useful Load: 1000 lbs
    Baggage: 120 lbs

1972

Model: 172L

Speed
    Top Speed at Sea Level: 139 mph
    Cruise, 75 percent power at 9000 ft: 131 mph
Range
    Cruise, 75 percent power at 9000 ft: 615 mi
        38 Gallons, No Reserve: 4.7 hrs at 131 mph
    Cruise, 75 percent at 9000 ft: 775 mi
        48 Gallons, No Reserve: 5.9 hrs at 131 mph
    Optimum Range at 10,000 ft: 640 mi
        38 Gallons, No Reserve: 5.5 hrs at 117 mph
    Optimum Range at 10,000 ft: 820 mi
        48 Gallons, No Reserve: 7.0 hrs at 117 mph

Performance
    Rate of Climb at Sea Level: 645 fpm
    Service Ceiling: 13,100 ft
    Takeoff
        Ground Run: 865 ft
        Over 50-ft Obstacle: 1525 ft
    Landing
        Ground Roll: 520 ft
        Over 50-ft Obstacle: 1250 ft
    Stall Speed
        Flaps Up, Power Off: 57 mph
        Flaps Down, Power Off: 49 mph
Capacities
    Standard Fuel: 42 gal
    w/optional tanks: 52 gal
    Oil: 8 qts
Engine
    Lycoming O-320-E2D
    TBO: 2000 hrs
    Power: (at 2700 rpm) 150 hp
    Propeller: (diameter) 75 in
Dimensions
    Wingspan: 35 ft 09 in
    Wing Area: (sq ft) 174
    Wing Loading: (lbs/sq ft) 13.2
    Power Loading: (lbs/hp) 15.3
    Length: 26 ft 11 in
    Height: 8 ft 09 in
Weights
    Gross: 2300 lbs
    Empty: 1265 lbs
    Useful Load: 1035 lbs
    Baggage: 120 lbs

Model: 172L Skyhawk

Speed
    Top Speed at Sea Level: 140 mph
    Cruise, 75 percent power at 9000 ft: 132 mph
Range
    Cruise, 75 percent power at 9000 ft: 620 mi
        38 Gallons, No Reserve: 4.7 hrs at 132 mph
    Cruise, 75 percent power at 9000 ft: 780 mi
        48 Gallons, No Reserve: 5.9 hrs at 132 mph
    Optimum Range at 10,000 ft: 655 mi
        38 Gallons, No Reserve: 5.5 hrs at 118 mph
    Optimum Range at 10,000 ft: 830 mi
        48 Gallons, No Reserve: 7.0 hrs at 118 mph

Performance
    Rate of Climb at Sea Level: 645 fpm
    Service Ceiling: 13,100 ft
    Takeoff
        Ground Run: 865 ft
        Over 50-ft Obstacle: 1525 ft
    Landing
        Ground Roll: 520 ft
        Over 50-ft Obstacle: 1250 ft
    Stall Speed
        Flaps Up, Power Off: 57 mph
        Flaps Down, Power Off: 49 mph
Capacities
    Standard Fuel: 42 gal
    w/optional tanks: 52 gal
    Oil: 8 qts
Engine
    Lycoming O-320-E2D
    TBO: 2000 hrs
    Power: (at 2700 rpm) 150 hp
    Propeller: (diameter) 75 in
Dimensions
    Wingspan: 35 ft 09 in
    Wing Area: (sq ft) 174
    Wing Loading: (lbs/sq ft) 13.2
    Power Loading: (lbs/hp) 15.3
    Length: 26 ft 11 in
    Height: 8 ft 09 in
Weights
    Gross: 2300 lbs
    Empty: 1305 lbs
    Useful Load: 995 lbs
    Baggage: 120 lbs

1973

Model: 172M

Speed
    Top Speed at Sea Level: 139 mph
    Cruise, 75 percent power at 9000 ft: 131 mph
Range
    Cruise, 75 percent power at 9000 ft: 615 mi
        38 Gallons, No Reserve: 4.7 hrs at 131 mph
    Cruise, 75 percent at 9000 ft: 775 mi
        48 Gallons, No Reserve: 5.9 hrs at 131 mph
    Optimum Range at 10,000 ft: 640 mi
        38 Gallons, No Reserve: 5.5 hrs at 117 mph

Optimum Range at 10,000 ft: 820 mi
     48 Gallons, No Reserve: 7.0 hrs at 117 mph

Performance
    Rate of Climb at Sea Level: 645 fpm
    Service Ceiling: 13,100 ft
    Takeoff
        Ground Run: 865 ft
        Over 50-ft Obstacle: 1525 ft
    Landing
        Ground Roll: 520 ft
        Over 50-ft Obstacle: 1250 ft
    Stall Speed
        Flaps Up, Power Off: 57 mph
        Flaps Down, Power Off: 51 mph

Capacities
    Standard Fuel: 42 gal
    w/optional tanks: 52 gal
    Oil: 8 qts

Engine
    Lycoming O-320-E2D
    TBO: 2000 hrs
    Power: (at 2700 rpm) 150 hp
    Propeller: (diameter) 75 in

Dimensions
    Wingspan: 35 ft 10 in
    Wing Area: (sq ft) 175.5
    Wing Loading: (lbs/sq ft) 13.2
    Power Loading: (lbs/hp) 15.3
    Length: 26 ft 11 in
    Height: 8 ft 09 in

Weights
    Gross: 2300 lbs
    Empty: 1285 lbs
    Useful Load: 1015 lbs
    Baggage: 120 lbs

Model: 172M Skyhawk

Speed
    Top Speed at Sea Level: 140 mph
    Cruise, 75 percent power at 9000 ft: 132 mph

Range
    Cruise, 75 percent power at 9000 ft: 620 mi
        38 Gallons, No Reserve: 4.7 hrs at 132 mph
    Cruise, 75 percent power at 9000 ft: 780 mi
        48 Gallons, No Reserve: 5.9 hrs at 132 mph
    Optimum Range at 10,000 ft: 655 mi
        38 Gallons, No Reserve: 5.5 hrs at 118 mph

Optimum Range at 10,000 ft: 830 mi
48 Gallons, No Reserve: 7.0 hrs at 118 mph

Performance
Rate of Climb at Sea Level: 645 fpm
Service Ceiling: 13,100 ft
Takeoff
Ground Run: 865 ft
Over 50-ft Obstacle: 1525 ft
Landing
Ground Roll: 520 ft
Over 50-ft Obstacle: 1250 ft
Stall Speed
Flaps Up, Power Off: 57 mph
Flaps Down, Power Off: 51 mph

Capacities
Standard Fuel: 42 gal
w/optional tanks: 52 gal
Oil: 8 qts

Engine
Lycoming O-320-E2D
TBO: 2000 hrs
Power: (at 2700 rpm) 150 hp
Propeller: (diameter) 75 in

Dimensions
Wingspan: 35 ft 10 in
Wing Area: (sq ft) 175.5
Wing Loading: (lbs/sq ft) 13.2
Power Loading: (lbs/hp) 15.3
Length: 26 ft 11 in
Height: 8 ft 09 in

Weights
Gross: 2300 lbs
Empty: 1335 lbs
Useful Load: 965 lbs
Baggage: 120 lbs

1974

Model: 172M

Speed
Top Speed at Sea Level: 140 mph
Cruise, 75 percent power at 8000 ft: 135 mph

Range
Cruise, 75 percent power at 8000 ft: 635 mi
38 Gallons, No Reserve: 4.7 hrs at 135 mph
Cruise, 75 percent at 8000 ft: 795 mi
48 Gallons, No Reserve: 5.9 hrs at 135 mph

Maximum Range at 10,000 ft: 695 mi
38 Gallons, No Reserve: 6.0 hrs at 116 mph
Maximum Range at 10,000 ft: 870 mi
48 Gallons, No Reserve: 7.5 hrs at 116 mph
Performance
Rate of Climb at Sea Level: 645 fpm
Service Ceiling: 13,100 ft
Takeoff
Ground Run: 865 ft
Over 50-ft Obstacle: 1525 ft
Landing
Ground Roll: 520 ft
Over 50-ft Obstacle: 1250 ft
Stall Speed
Flaps Up, Power Off: 57 mph
Flaps Down, Power Off: 49 mph
Capacities
Standard Fuel: 42 gal
w/optional tanks: 52 gal
Oil: 8 qts
Engine
Lycoming O-320-E2D
TBO: 2000 hrs
Power: (at 2700 rpm) 150 hp
Propeller: (diameter) 75 in
Dimensions
Wingspan: 35 ft 10 in
Wing Area: (sq ft) 174
Wing Loading: (lbs/sq ft) 13.2
Power Loading: (lbs/hp) 15.3
Length: 26 ft 11 in
Height: 8 ft 09 in
Weights
Gross: 2300 lbs
Empty: 1300 lbs
Useful Load: 1000 lbs
Baggage: 120 lbs

Model: 172M Skyhawk and Skyhawk II

Speed
Top Speed at Sea Level: 144 mph
Cruise, 75 percent power at 8000 ft: 138 mph
Range
Cruise, 75 percent power at 8000 ft: 650 mi
38 Gallons, No Reserve: 4.7 hrs at 138 mph
Cruise, 75 percent power at 8000 ft: 815 mi
48 Gallons, No Reserve: 5.9 hrs at 138 mph

Maximum Range at 10,000 ft: 700 mi
38 Gallons, No Reserve: 6.0 hrs at 117 mph
Maximum Range at 10,000 ft: 875 mi
48 Gallons, No Reserve: 7.5 hrs at 117 mph

Performance
Rate of Climb at Sea Level: 645 fpm
Service Ceiling: 13,100 ft
Takeoff
Ground Run: 865 ft
Over 50-ft Obstacle: 1525 ft
Landing
Ground Roll: 520 ft
Over 50-ft Obstacle: 1250 ft
Stall Speed
Flaps Up, Power Off: 57 mph
Flaps Down, Power Off: 49 mph

Capacities
Standard Fuel: 42 gal
w/optional tanks: 52 gal
Oil: 8 qts

Engine
Lycoming O-320-E2D
TBO: 2000 hrs
Power: (at 2700 rpm) 150 hp
Propeller: (diameter) 75 in

Dimensions
Wingspan: 35 ft 10 in
Wing Area: (sq ft) 174
Wing Loading: (lbs/sq ft) 13.2
Power Loading: (lbs/hp) 15.3
Length: 26 ft 11 in
Height: 8 ft 09 in

Weights
Gross: 2300 lbs
Empty
(Skyhawk): 1345 lbs
(Skyhawk II): 1370 lbs
Useful Load
(Skyhawk): 955 lbs
(Skyhawk II): 930 lbs
Baggage: 120 lbs

1975

Model: 172M

Speed
Top Speed at Sea Level: 140 mph

Cruise, 75 percent power at 8000 ft: 135 mph
Range
    Cruise, 75 percent power at 8000 ft: 635 mi
        38 Gallons, No Reserve: 4.7 hrs at 135 mph
    Cruise, 75 percent at 8000 ft: 795 mi
        48 Gallons, No Reserve: 5.9 hrs at 135 mph
    Maximum Range at 10,000 ft: 695 mi
        38 Gallons, No Reserve: 6.0 hrs at 116 mph
    Maximum Range at 10,000 ft: 870 mi
        48 Gallons, No Reserve: 7.5 hrs at 116 mph
Performance
    Rate of Climb at Sea Level: 645 fpm
    Service Ceiling: 13,100 ft
    Takeoff
        Ground Run: 865 ft
        Over 50-ft Obstacle: 1525 ft
    Landing
        Ground Roll: 520 ft
        Over 50-ft Obstacle: 1250 ft
    Stall Speed
        Flaps Up, Power Off: 57 mph
        Flaps Down, Power Off: 49 mph
Capacities
    Standard Fuel: 42 gal
    w/optional tanks: 52 gal
    Oil: 8 qts
Engine
    Lycoming O-320-E2D
    TBO: 2000 hrs
    Power: (at 2700 rpm) 150 hp
    Propeller: (diameter) 75 in
Dimensions
    Wingspan: 35 ft 10 in
    Wing Area: (sq ft) 174
    Wing Loading: (lbs/sq ft) 13.2
    Power Loading: (lbs/hp) 15.3
    Length: 26 ft 11 in
    Height: 8 ft 09 in
Weights
    Gross: 2300 lbs
    Empty: 1300 lbs
    Useful Load: 1000 lbs
    Baggage: 120 lbs

Model: 172M Skyhawk and Skyhawk II

Speed
    Top Speed at Sea Level: 144 mph
    Cruise, 75 percent power at 8000 ft: 138 mph

Range
    Cruise, 75 percent power at 8000 ft: 650 mi
        38 Gallons, No Reserve: 4.7 hrs at 138 mph
    Cruise, 75 percent power at 8000 ft: 815 mi
        48 Gallons, No Reserve: 5.9 hrs at 138 mph
    Maximum Range at 10,000 ft: 700 mi
        38 Gallons, No Reserve: 6.0 hrs at 117 mph
    Maximum Range at 10,000 ft: 875 mi
        48 Gallons, No Reserve: 7.5 hrs at 117 mph

Performance
    Rate of Climb at Sea Level: 645 fpm
    Service Ceiling: 13,100 ft
    Takeoff
        Ground Run: 865 ft
        Over 50-ft Obstacle: 1525 ft
    Landing
        Ground Roll: 520 ft
        Over 50-ft Obstacle: 1250 ft
    Stall Speed
        Flaps Up, Power Off: 57 mph
        Flaps Down, Power Off: 49 mph

Capacities
    Standard Fuel: 42 gal
    w/optional tanks: 52 gal
    Oil: 8 qts

Engine
    Lycoming O-320-E2D
    TBO: 2000 hrs
    Power: (at 2700 rpm) 150 hp
    Propeller: (diameter) 75 in

Dimensions
    Wingspan: 35 ft 10 in
    Wing Area: (sq ft) 174
    Wing Loading: (lbs/sq ft) 13.2
    Power Loading: (lbs/hp) 15.3
    Length: 26 ft 11 in
    Height: 8 ft 09 in

Weights
    Gross: 2300 lbs
    Empty
        (Standard): 1305 lbs
        (Skyhawk): 1345 lbs
        (Skyhawk II): 1370 lbs
    Useful Load
        (Standard): 995 lbs
        (Skyhawk): 955 lbs
        (Skyhawk II): 930 lbs
    Baggage: 120 lbs

1976
Model: 172M Skyhawk and Skyhawk II

Speed
    Top Speed at Sea Level: 125 kts
    Cruise, 75 percent power at 8000 ft: 120 kts
Range (with 45 minute reserve)
    Cruise, 75 percent power at 8000 ft: 450 nm
        38 Gallons Usable Fuel: 3.9 hrs
    Cruise, 75 percent at 8000 ft: 595 nm
        48 Gallons Usable Fuel: 5.1 hrs
    Maximum Range at 10,000 ft: 480 nm
        38 Gallons Usable Fuel: 4.8 hrs
    Maximum Range at 10,000 ft: 640 nm
        48 Gallons Usable Fuel: 6.3 hrs
Performance
    Rate of Climb at Sea Level: 645 fpm
    Service Ceiling: 13,100 ft
    Takeoff
        Ground Run: 865 ft
        Over 50-ft Obstacle: 1525 ft
    Landing
        Ground Roll: 520 ft
        Over 50-ft Obstacle: 1250 ft
    Stall Speed
        Flaps Up, Power Off: 42 kts
        Flaps Down, Power Off: 36 kts
Capacities
    Standard Fuel: 42 gal
    w/optional tanks: 52 gal
    Oil: 8 qts
Engine
    Lycoming O-320-E2D
    TBO: 2000 hrs
    Power: (at 2700 rpm) 150 hp
    Propeller: (diameter) 75 in
Dimensions
    Wingspan: 35 ft 10 in
    Wing Area: (sq ft) 174
    Wing Loading: (lbs/sq ft) 13.2
    Power Loading: (lbs/hp) 15.3
    Length: 26 ft 11 in
    Height: 8 ft 09 in
Weights
    Maximum: 2300 lbs
    Empty
        (Skyhawk): 1387 lbs

(Skyhawk II): 1412 lbs
Useful Load
    (Skyhawk): 913 lbs
    (Skyhawk II): 888 lbs
Baggage: 120 lbs

1977

Model: 172N Skyhawk/100 and Skyhawk II/100

Speed
    Top Speed at Sea Level: 125 kts
    Cruise, 75 percent power at 8000 ft: 122 kts
Range (with 45 minute reserve)
    Cruise, 75 percent power at 8000 ft: 485 nm
        40 Gallons Usable Fuel: 4.1 hrs
    Cruise, 75 percent at 8000 ft: 630 nm
        50 Gallons Usable Fuel: 5.3 hrs
    Maximum Range at 10,000 ft: 575 nm
        40 Gallons Usable Fuel: 5.7 hrs
    Maximum Range at 10,000 ft: 750 nm
        50 Gallons Usable Fuel: 7.4 hrs
Performance
    Rate of Climb at Sea Level: 770 fpm
    Service Ceiling: 14,200 ft
    Takeoff
        Ground Run: 820 ft
        Over 50-ft Obstacle: 1440 ft
    Landing
        Ground Roll: 520 ft
        Over 50-ft Obstacle: 1250 ft
    Stall Speed
        Flaps Up, Power Off: 50 kts
        Flaps Down, Power Off: 44 kts
Capacities
    Standard Fuel: 43 gal
    w/optional tanks: 54 gal
    Oil: 6 qts
Engine
    Lycoming O-320-H2AD
    TBO: 2000 hrs
    Power: (at 2700 rpm) 160 hp
    Propeller: (diameter) 75 in
Dimensions
    Wingspan: 35 ft 10 in
    Wing Area: (sq ft) 174
    Wing Loading: (lbs/sq ft) 13.2

Power Loading: (lbs/hp) 14.4
Length: 26 ft 11 in
Height: 8 ft 09 in
Weights
Maximum: 2300 lbs
Empty
(Skyhawk): 1379 lbs
(Skyhawk II): 1403 lbs
Useful Load
(Skyhawk): 921 lbs
(Skyhawk II): 897 lbs
Baggage: 120 lbs

1978

Model: 172N Skyhawk/100 and Skyhawk II/100

Speed
Top Speed at Sea Level: 125 kts
Cruise, 75 percent power at 8000 ft: 122 kts
Range (with 45 minute reserve)
Cruise, 75 percent power at 8000 ft: 485 nm
40 Gallons Usable Fuel: 4.1 hrs
Cruise, 75 percent at 8000 ft: 630 nm
50 Gallons Usable Fuel: 5.3 hrs
Maximum Range at 10,000 ft: 575 nm
40 Gallons Usable Fuel: 5.7 hrs
Maximum Range at 10,000 ft: 750 nm
50 Gallons Usable Fuel: 7.4 hrs
Performance
Rate of Climb at Sea Level: 770 fpm
Service Ceiling: 14,200 ft
Takeoff
Ground Run: 805 ft
Over 50-ft Obstacle: 1440 ft
Landing
Ground Roll: 520 ft
Over 50-ft Obstacle: 1250 ft
Stall Speed
Flaps Up, Power Off: 50 kts
Flaps Down, Power Off: 44 kts
Capacities
Standard Fuel: 43 gal
w/optional tanks: 54 gal
Oil: 6 qts
Engine
Lycoming O-320-H2AD
TBO: 2000 hrs

Power: (at 2700 rpm) 160 hp
Propeller: (diameter) 75 in
Dimensions
Wingspan: 35 ft 10 in
Wing Area: (sq ft) 174
Wing Loading: (lbs/sq ft) 13.2
Power Loading: (lbs/hp) 14.4
Length: 26 ft 11 in
Height: 8 ft 09 in
Weights
Gross: 2300 lbs
Empty
(Skyhawk): 1393 lbs
(Skyhawk II): 1419 lbs
Useful Load
(Skyhawk): 907 lbs
(Skyhawk II): 881 lbs
Baggage: 120 lbs

1979
Model: 172N Skyhawk/100 and Skyhawk II/100

Speed
Top Speed at Sea Level: 125 kts
Cruise, 75 percent power at 8000 ft: 122 kts
Range (with 45 minute reserve)
Cruise, 75 percent power at 8000 ft: 485 nm
40 Gallons Usable Fuel: 4.1 hrs
Cruise, 75 percent at 8000 ft: 630 nm
50 Gallons Usable Fuel: 5.3 hrs
Maximum Range at 10,000 ft: 575 nm
40 Gallons Usable Fuel: 5.7 hrs
Maximum Range at 10,000 ft: 750 nm
50 Gallons Usable Fuel: 7.4 hrs
Performance
Rate of Climb at Sea Level: 770 fpm
Service Ceiling: 14,200 ft
Takeoff
Ground Run: 805 ft
Over 50-ft Obstacle: 1440 ft
Landing
Ground Roll: 520 ft
Over 50-ft Obstacle: 1250 ft
Stall Speed
Flaps Up, Power Off: 50 kts
Flaps Down, Power Off: 44 kts

Capacities
    Standard Fuel: 43 gal
    w/optional tanks: 54 gal
    Oil: 6 qts
Engine
    Lycoming O-320-H2AD
    TBO: 2000 hrs
    Power: (at 2700 rpm) 160 hp
    Propeller: (diameter) 75 in
Dimensions
    Wingspan: 35 ft 10 in
    Wing Area: (sq ft) 174
    Wing Loading: (lbs/sq ft) 13.2
    Power Loading: (lbs/hp) 14.4
    Length: 26 ft 11 in
    Height: 8 ft 09 in
Weights
    Maximum: 2300 lbs
    Empty
        (Skyhawk): 1379 lbs
        (Skyhawk II): 1424 lbs
    Useful Load
        (Skyhawk): 910 lbs
        (Skyhawk II): 983 lbs
    Baggage: 120 lbs

1980

Model: 172N Skyhawk/100 and Skyhawk II/100

Speed
    Top Speed at Sea Level: 125 kts
    Cruise, 75 percent power at 8000 ft: 122 kts
Range (with 45 minute reserve)
    Cruise, 75 percent power at 8000 ft: 455 nm
        40 Gallons Usable Fuel: 3.8 hrs
    Cruise, 75 percent at 8000 ft: 600 nm
        50 Gallons Usable Fuel: 5.0 hrs
    Maximum Range at 10,000 ft: 575 nm
        40 Gallons Usable Fuel: 6.1 hrs
    Maximum Range at 10,000 ft: 750 nm
        50 Gallons Usable Fuel: 7.9 hrs
Performance
    Rate of Climb at Sea Level: 770 fpm
    Service Ceiling: 14,200 ft
    Takeoff
        Ground Run: 775 ft
        Over 50-ft Obstacle: 1390 ft

Landing
  Ground Roll: 520 ft
  Over 50-ft Obstacle: 1250 ft
Stall Speed
  Flaps Up, Power Off: 50 kts
  Flaps Down, Power Off: 44 kts
Capacities
  Standard Fuel: 43 gal
  w/optional tanks: 54 gal
  Oil: 6 qts
Engine
  Lycoming O-320-H2AD
  TBO: 2000 hrs
  Power: (at 2700 rpm) 160 hp
  Propeller: (diameter) 75 in
Dimensions
  Wingspan: 35 ft 10 in
  Wing Area: (sq ft) 174
  Wing Loading: (lbs/sq ft) 13.2
  Power Loading: (lbs/hp) 14.4
  Length: 26 ft 11 in
  Height: 8 ft 09 in
Weights
  Maximum: 2300 lbs
  Empty
    (Skyhawk): 1403 lbs
    (Skyhawk II): 1430 lbs
  Useful Load
    (Skyhawk): 904 lbs
    (Skyhawk II): 877 lbs
  Baggage: 120 lbs

1981

Model: 172P Skyhawk and Skyhawk II

Speed
  Top Speed at Sea Level: 123 kts
  Cruise, 75 percent power at 8000 ft: 120 kts
Range (with 45 minute reserve)
  Cruise, 75 percent power at 8000 ft: 440 nm
    40 Gallons Usable Fuel: 3.8 hrs
  Cruise, 75 percent at 8000 ft: 585 nm
    50 Gallons Usable Fuel: 5.0 hrs
  Cruise, 75 percent at 8000 ft: 755 nm
    60 Gallons Usable Fuel: 6.4 hrs
  Maximum Range at 10,000 ft: 520 nm
    40 Gallons Usable Fuel: 5.6 hrs

Maximum Range at 10,000 ft: 680 nm
　　50 Gallons Usable Fuel: 7.4 hrs
Maximum Range at 10,000 ft: 875 nm
　　60 Gallons Usable Fuel: 9.4 hrs
Performance
　　Rate of Climb at Sea Level: 700 fpm
　　Service Ceiling: 13,000 ft
　　Takeoff
　　　　Ground Run: 890 ft
　　　　Over 50-ft Obstacle: 1825 ft
　　Landing
　　　　Ground Roll: 540 ft
　　　　Over 50-ft Obstacle: 1280 ft
　　Stall Speed
　　　　Flaps Up, Power Off: 51 kts
　　　　Flaps Down, Power Off: 46 kts
Capacities
　　Standard Fuel: 43 gal
　　w/long range tanks: 54 gal
　　w/integral long range tanks: 68 gal
　　Oil: 8 qts
Engine
　　Lycoming O-320-D2J
　　TBO: 2000 hrs
　　Power: (at 2700 rpm) 160 hp
　　Propeller: (diameter) 75 in
Dimensions
　　Wingspan: 35 ft 10 in
　　Wing Area: (sq ft) 174
　　Wing Loading: (lbs/sq ft) 13.8
　　Power Loading: (lbs/hp) 15.0
　　Length: 26 ft 11 in
　　Height: 8 ft 09 in
Weights
　　Maximum: 2400 lbs
　　Empty
　　　　(Skyhawk): 1411 lbs
　　　　(Skyhawk II): 1439 lbs
　　Useful Load
　　　　(Skyhawk): 996 lbs
　　　　(Skyhawk II): 968 lbs
　　Baggage: 120 lbs

1982

Model: 172P Skyhawk and Skyhawk II

Speed
　　Top Speed at Sea Level: 123 kts

Cruise, 75 percent power at 8000 ft: 120 kts
Range (with 45 minute reserve)
    Cruise, 75 percent power at 8000 ft: 440 nm
        40 Gallons Usable Fuel: 3.8 hrs
    Cruise, 75 percent at 8000 ft: 585 nm
        50 Gallons Usable Fuel: 5.0 hrs
    Cruise, 75 percent at 8000 ft: 755 nm
        60 Gallons Usable Fuel: 6.4 hrs
    Maximum Range at 10,000 ft: 520 nm
        40 Gallons Usable Fuel: 5.6 hrs
    Maximum Range at 10,000 ft: 680 nm
        50 Gallons Usable Fuel: 7.4 hrs
    Maximum Range at 10,000 ft: 875 nm
        60 Gallons Usable Fuel: 9.4 hrs

Performance
    Rate of Climb at Sea Level: 700 fpm
    Service Ceiling: 13,000 ft
    Takeoff
        Ground Run: 890 ft
        Over 50-ft Obstacle: 1625 ft
    Landing
        Ground Roll: 540 ft
        Over 50-ft Obstacle: 1280 ft
    Stall Speed
        Flaps Up, Power Off: 51 kts
        Flaps Down, Power Off: 46 kts

Capacities
    Standard Fuel: 43 gal
        w/long range tanks: 54 gal
        w/integral long range tanks: 68 gal
    Oil: 8 qts

Engine
    Lycoming O-320-D2J
    TBO: 2000 hrs
    Power: (at 2700 rpm) 160 hp
    Propeller: (diameter) 75 in

Dimensions
    Wingspan: 35 ft 10 in
    Wing Area: (sq ft) 174
    Wing Loading: (lbs/sq ft) 13.8
    Power Loading: (lbs/hp) 15.0
    Length: 26 ft 11 in
    Height: 8 ft 09 in

Weights
    Maximum: 2400 lbs
    Empty
        (Skyhawk): 1430 lbs
        (Skyhawk II): 1448 lbs

Useful Load
    (Skyhawk): 977 lbs
    (Skyhawk II): 959 lbs
Baggage: 120 lbs

1983
Model: 172 Skyhawk and Skyhawk II

Speed
    Top Speed at Sea Level: 123 kts
    Cruise, 75 percent power at 8000 ft: 120 kts
Range (with 45 minute reserve)
    Cruise, 75 percent power at 8000 ft: 440 nm
        40 Gallons Usable Fuel: 3.8 hrs
    Cruise, 75 percent at 8000 ft: 585 nm
        50 Gallons Usable Fuel: 5.0 hrs
    Cruise, 75 percent at 8000 ft: 755 nm
        62 Gallons Usable Fuel: 6.4 hrs
    Maximum Range at 10,000 ft: 520 nm
        40 Gallons Usable Fuel: 5.6 hrs
    Maximum Range at 10,000 ft: 680 nm
        50 Gallons Usable Fuel: 7.4 hrs
    Maximum Range at 10,000 ft: 875 nm
        62 Gallons Usable Fuel: 9.4 hrs
Performance
    Rate of Climb at Sea Level: 700 fpm
    Service Ceiling: 13,000 ft
    Takeoff
        Ground Run: 890 ft
        Over 50-ft Obstacle: 1625 ft
    Landing
        Ground Roll: 540 ft
        Over 50-ft Obstacle: 1280 ft
    Stall Speed
        Flaps Up, Power Off: 51 kts
        Flaps Down, Power Off: 46 kts
Capacities
    Standard Fuel: 43 gal
    w/long range tanks: 54 gal
    w/integral long range tanks: 68 gal
    Oil: 8 qts
Engine
    Lycoming O-320-D2J
    TBO: 2000 hrs
    Power: (at 2700 rpm) 160 hp
    Propeller: (diameter) 75 in
Dimensions

Wingspan: 35 ft 10 in
Wing Area: (sq ft) 174
Wing Loading: (lbs/sq ft) 13.8
Power Loading: (lbs/hp) 15.0
Length: 26 ft 11 in
Height: 8 ft 09 in
Weights
Maximum: 2400 lbs
Empty
(Skyhawk): 1439 lbs
(Skyhawk II): 1466 lbs
Useful Load
(Skyhawk): 968 lbs
(Skyhawk II): 941 lbs
Baggage: 120 lbs

1984

Model: 172 Skyhawk and Skyhawk II

Speed
Top Speed at Sea Level: 123 kts
Cruise, 75 percent power: 120 kts
Range
75 percent at 8000 ft
40 Gallons, 45 min. Reserve: 440 nm/3.8 hrs
50 Gallons, 45 min. Reserve: 585 nm/5.0 hrs
62 Gallons, 45 min. Reserve: 775 nm/6.4 hrs
Maximum Range at 10,000 ft
40 Gallons 520 nm/5.6 hrs
50 Gallons 680 nm/7.4 hrs
62 Gallons 875 nm/9.4 hrs
Performance
Rate of Climb at Sea Level: 700 fpm
Service Ceiling: 13,000 ft
Takeoff
Ground Run: 890 ft
Over 50-ft Obstacle: 1625 ft
Landing
Ground Roll: 540 ft
Over 50-ft Obstacle: 1280 ft
Stall Speed
Flaps Up, Power Off: 51 kts
Flaps Down, Power Off: 46 kts
Capacities
Standard Fuel: 43 gal
Long range tanks: 54 gal
Integral long range tanks: 68 gal

Engine
    Lycoming O-320-D2J
    TBO: 2000 hrs
    Power: (at 2700 rpm) 160 hp
    Propeller: Fixed Pitch
Dimensions
    Wingspan: 36 ft 01 in
    Wing Area: (sq ft) 174
    Length: 26 ft 11 in
    Height: 8 ft 09 in
Weights
    Gross: 2400 lbs
    Empty
        (Skyhawk): 1438 lbs
        (Skyhawk II): 1457 lbs
    Useful Load
        (Skyhawk): 969 lbs
        (Skyhawk II): 950 lbs
        Baggage: 120 lbs

1985
Model: 172 Skyhawk and Skyhawk II

Speed
    Top Speed at Sea Level: 123 kts
    Cruise, 75 percent power: 120 kts
Range
    75 percent at 8000 ft
        40 Gallons, 45 min. Reserve: 440 nm/3.8 hrs
        50 Gallons, 45 min. Reserve: 585 nm/5.0 hrs
        62 Gallons, 45 min. Reserve: 775 nm/6.4 hrs
    Maximum Range at 10,000 ft
        40 Gallons 520 nm/5.6 hrs
        50 Gallons 680 nm/7.4 hrs
        62 Gallons 875 nm/9.4 hrs
Performance
    Rate of Climb at Sea Level: 700 fpm
    Service Ceiling: 13,000 ft
    Takeoff
        Ground Run: 890 ft
        Over 50-ft Obstacle: 1625 ft
    Landing
        Ground Roll: 540 ft
        Over 50-ft Obstacle: 1280 ft
    Stall Speed
        Flaps Up, Power Off: 51 kts
        Flaps Down, Power Off: 46 kts

Capacities
  Standard Fuel: 43 gal
  Long range tanks: 54 gal
  Integral long range tanks: 68 gal
Engine
  Lycoming O-320-D2J
  TBO: 2000 hrs
  Power: (at 2700 rpm) 160 hp
  Propeller: Fixed Pitch
Dimensions
  Wingspan: 36 ft 01 in
  Wing Area: (sq ft) 174
  Length: 26 ft 11 in
  Height: 8 ft 09 in
Weights
  Gross: 2407 lbs
  Empty
    (Skyhawk): 1433 lbs
    (Skyhawk II): 1452 lbs
  Useful Load
    (Skyhawk): 974 lbs
    (Skyhawk II): 955 lbs
  Baggage: 120 lbs

1986
Model: 172 Skyhawk and Skyhawk II

Speed
  Top Speed at Sea Level: 123 kts
  Cruise, 75 percent power: 120 kts
Range
  75 percent at 8000 ft
    40 Gallons, 45 min. Reserve: 440 nm/3.8 hrs
    50 Gallons, 45 min. Reserve: 585 nm/5.0 hrs
    62 Gallons, 45 min. Reserve: 775 nm/6.4 hrs
  Maximum Range at 10,000 ft
    40 Gallons 520 nm/5.6 hrs
    50 Gallons 680 nm/7.4 hrs
    62 Gallons 875 nm/9.4 hrs
Performance
  Rate of Climb at Sea Level: 700 fpm
  Service Ceiling: 13,000 ft
  Takeoff
    Ground Run: 890 ft
    Over 50-ft Obstacle: 1625 ft
  Landing
    Ground Roll: 540 ft

Over 50-ft Obstacle: 1280 ft
Stall Speed
Flaps Up, Power Off: 51 kts
Flaps Down, Power Off: 46 kts
Capacities
Standard Fuel: 43 gal
Long range tanks: 54 gal
Integral long range tanks: 68 gal
Engine
Lycoming O-320-D2J
TBO: 2000 hrs
Power: (at 2700 rpm) 160 hp
Propeller: Fixed Pitch
Dimensions
Wingspan: 36 ft 01 in
Wing Area: (sq ft) 174
Length: 26 ft 11 in
Height: 8 ft 09 in
Weights
Gross: 2407 lbs
Empty
(Skyhawk): 1433 lbs
(Skyhawk II): 1452 lbs
Useful Load
(Skyhawk): 974 lbs
(Skyhawk II): 955 lbs
Baggage: 120 lbs

3

Engines

ONLY TWO POWERPLANTS WERE INSTALLED on basic 172s: the Continental O-300 and the Lycoming O-320. Many variations of the two engines are on 172s. Here is a short list of the original factory-installed engines:

YEAR	ENGINE	HP	TBO
1956–59	Cont O-300-A	145	1800
1960	Cont O-300-C	145	1800
1961–67	Cont O-300-D	145	1800
1968–76	Lyco O-320-E2D	150	2000
1977–80	Lyco O-320-H2AD	160	2000
1981–86	Lyco O-320-D2J	160	2000

VARIANT ENGINES

Other factory-installed engines are found in the variant 172s: 175, Hawk XP, Cutlass RG, and Cutlass.

Model 175

YEAR	ENGINE	HP	TBO
1958–59	Cont GO-300-A	175	1200
1960	Cont GO-300-C	175	1200
1961	Cont GO-300-D	175	1200
1962	Cont GO-300-E	175	1200

Model 172 Powermatic

YEAR	ENGINE	HP	TBO
1963	Cont GO-300-E	175	1200

Model 172 Hawk XP

YEAR	ENGINE	HP	TBO
1977–78	Cont IO-360-K	195	1500
1979–81	Cont IO-360-KB	195	2000

Model 172 Cutlass RG

YEAR	ENGINE	HP	TBO
1980–85	Lyco O-360-F1A6	180	2000

Model 172 Cutlass

YEAR	ENGINE	HP	TBO
1983–84	Lyco O-360-A4N	180	2000

CONTINENTAL ENGINES

Continental engines used in the early 172s are long out of production. This does not mean they are not good engines, just that they are no longer made. This should be a consideration when making a purchase because no longer in production means dwindling parts supplies and higher costs for repairs.

All the Continental engines listed for the 172 series airplanes are six-cylinder design (Fig. 3-1). The O-300 engine in the 172 and the O-200 engine in the Cessna Model 150 are very close cousins. Many small parts are interchangeable between the O-200 and O-300 engines. The O-300 engines all operate on 80/87 octane avgas. The IO-360 requires 100/130 octane.

Fig. 3-1. The Continental O-300 engine that powered the early Cessna 172s. Teledyne Continental

The GO-300 engines in the Model 175s have gear reduction units to reduce the rpm of the propeller as compared to the crankshaft. The engine turns at 3200 rpm to develop 175 hp. Propeller tips would go supersonic at 3200 rpm; hence, the gear reduction unit.

Later IO-360 engines in the Hawk XP are fuel-injected and are in current production. These engines are more complex than the predecessors and are more expensive to maintain and, when the time comes, to rebuild.

LYCOMING ENGINES

Three basic Lycoming engines found in the 172s are all O-320 models. The first is the O-320-E2D, which was introduced in 1968 and remained through 1976 (Fig. 3-2). This engine is among the most reliable workhorse aircraft powerplants ever produced. It operates on 80/87 avgas.

Fig. 3-2. The Lycoming O-320 engine used in the later Cessna 172s. Avco Lycoming

In 1977, the O-320-H2AD became the standard engine for the 172s. It boasted an extra 10 horsepower and was designed to utilize 100LL avgas because 80-octane was becoming scarce. This engine proved to be a complete disaster. Many self-destructed long before the recommended 2000-hour TBO. Catastrophic failure equates to total engine failure, sometimes leaving a trail of oil and parts, and resulting in an unplanned landing. Several airworthiness directives have been issued for the engine. Numerous "fixes" keep these engines running. For more information about the O-320-H2AD engine, *see* chapter 4 and chapter 12.

Cessna replaced the H2AD engine with the O-320-D2J engine in 1981.

ENGINE SPECIFICATIONS

Continental O-300
Horsepower: 145 at 2700 rpm
Number of Cylinders: 6 Horizontally Opposed
    Displacement: 301.37 cu in
    Bore: 4.0625 in
    Stroke: 3.875 in
    Compression Ratio: 7.0:1
Magnetos: Slick 664
    Right: Fires 26 degrees BTC upper
    Left: Fires 28 degrees BTC lower
Firing Order: 1-6-3-2-5-4
Spark Plugs: SH20A
    Gap: .018 to .022 in
Torque: 330 lbs-in
Carburetor: Marvel-Schebler MA-3-SPA
Alternator: 14 volts at 60 amps
Starter: Automatic engagement
Tachometer: mechanical
Oil Capacity: 8 qts
Oil Pressure
    Minimum at idle: 5 psi
    Normal: 30–60 psi
    Maximum (start-up): 100 psi
Propeller Rotation: Clockwise (viewed from rear)
Dry Weight: 298 lbs

Lycoming O-320
Horsepower
    (E2D) 150 at 2700 rpm
    (H2AD) 160 at 2700 rpm
    (D2J) 160 at 2700 rpm
Number of Cylinders: 4 Horizontally Opposed
    Displacement: 319.8 cu. in.
    Bore: 5.125 in.
    Stroke: 3.875 in.
    Compression Ratio
     (E2D) 7.0:1
     (H2AD) 9.0:1
     (D2J) 8.5:1
Magnetos: Slick 4051 (left) 4050 (right)
    Right: Fires 25 degrees BTC 1-3 upper and 2-4 lower
    Left: Fires 25 degrees BTC 2-4 upper and 1-3 lower
    Note: H2AD has D2RN-2021 impulse coupling dual magneto and the D2J has two Slick magnetos.
    Firing Order: 1-3-2-4
Spark Plugs: SH15

Gap: .015 to .018 in
Torque: 390 lbs-in
Carburetor: Marvel-Schebler MA-4SPA
Alternator:14 volts at 60 amps (28 volts after 1977)
Starter: Automatic engagement
Tachometer: Mechanical
Oil Capacity: 8 qts
Oil Pressure
Minimum at idle: 25 psi
Normal: 60–90 psi
Maximum (start-up): 100 psi
Propeller Rotation: Clockwise (viewed from rear)
Dry Weight: 269 lbs

ENGINE LANGUAGE

The more you know about engines, the better. Here are a few definitions to help you understand them better:

TBO. The time between overhauls recommended by the manufacturer as the maximum engine life. It has no legal bearing on airplanes not used in commercial service; it's only an indicator. Many well-cared-for engines last hundreds of hours beyond TBO, but not all.

Overhaul. Disassembly, repair, inspection, cleaning, and reassembly of an engine. There is no FAR standard; therefore, the work may be done to new limits or to service limits.

Rebuild. Disassembly, repair, alteration, inspection, cleaning and reassembly of an engine, including bringing all specifications back to factory-new limits. In accordance with the FARs, only the engine's manufacturer can rebuild an engine. When rebuilt, the engine is zero timed and comes with a new logbook.

Remanufactured. A term having no official validity, but often used by engine overhaul shops. The term generally equates to an overhaul to new limits.

Top overhaul. Rebuilding of the head assemblies, but not of the entire engine. In other words, the case of the engine is not split, only the cylinders are pulled. Top overhaul is utilized to bring engines that burn oil or have low compressions within specifications. It is a method of stretching the life of an otherwise sound engine. A top overhaul can include such work as valve replacement or grinding, cylinder replacement or repair, piston and ring replacement, and the like. It is not necessarily an indicator of a poor engine. The need for a top overhaul might have been brought on by such things as pilot abuse, lack of care, lack of use, or abuse (hard climbs and fast letdowns). Note that the term top overhaul does not indicate the extent of the rebuild job—number of cylinders rebuilt or the completeness of the job.

New limits. Dimensions and specifications used when constructing a new engine. Parts meeting new limits will normally reach TBO with no further attention, save for routine maintenance.

Service limits. Dimensions and specifications below which use is forbidden. Many used engine parts will fit into this category; however, they are unlikely to last the full TBO, as they are already partially worn.

Magnaflux and magnaglow. Methods of detecting invisible defects (cracks) in ferrous metals. Parts normally magnafluxed or magnaglowed are crankshafts, camshafts, piston pins, rocker arms, and the like. Magnaflux and magnaglow inspections are routinely done during remanufacture and overhaul.

Nitriding. A means of hardening cylinder barrels and crankshafts. The purpose is to create a hard surface that resists wear, thereby extending the useful life of the part.

Chromeplating. Used to bring the internal dimensions of the cylinders back to specifications. The plating produces a hard, machinable, and long-lasting surface. The major drawback of chromeplating is a longer break-in time; however, an advantage of the chromeplating is a resistance to destructive oxidation (rust) within combustion chambers.

Cermicrome. Trademarked process of chromeplating combined with an oil wettable silicon carbide impregnated coating.

Cylinder color codes. When looking at the engine of an airplane you can sometimes see a little of the past work done on it. You will notice that some of the cylinders (often referred to as *jugs*) might be painted or banded. The colors of the paint or band tell you about the physical properties of the individual cylinder:

- Orange indicates a chromeplated cylinder barrel
- Blue indicates a nitrided cylinder barrel
- Green indicates internal cylinder dimension is .010 oversize
- Yellow indicates .020 oversize

USED ENGINES

Many airplane ads proudly state the hours on the engine (716 SMOH). Basically, this means that there have been 716 hours of operation since the engine was overhauled. Not stated is how it was used, or how completely it was overhauled. The time on an engine, since new or overhaul, is an important factor when placing a value on an airplane. The recommended TBO, less the hours currently on the engine, is the time remaining. The difference between these times is the expected remaining life of the engine.

Three basic terms are normally used when referring to the time on an airplane engine:

- Low time—first ⅓ of TBO
- Mid time—second ⅓ of TBO
- High time—last ⅓ of TBO

Naturally, other variables come into play when referring to TBO:

- Are the hours on the engine since new, remanufacture, or overhaul?
- What type of flying has the engine seen?
- Was it flown on a regular basis?
- What kind of maintenance did the engine get?

The engine logbook will indicate whether the engine is operating on time since new, remanufacture, or overhaul. The logs should also be of some help in determining questions about engine maintenance. Preventive maintenance should have been accomplished and logged throughout the engine's life (oil changes, plug changes, and the like). In accordance with FARs, all maintenance must be logged.

Airplanes that have not been flown on a regular basis—and maintained in a similar fashion—will never reach full TBO. Manufacturers refer to regular usage as 20–40 hours monthly: however, there are few privately owned airplanes meeting the upper limits of this requirement. Most pilots don't have the time or money required for such constant use. Logging 20-to-40 hours monthly equates to 240 to 480 hours yearly. When an engine isn't run, acids and moisture in the oil will oxidize engine components. In addition, the lack of lubricant movement will cause the seals to dry out. Left long enough, the engine will seize and no longer be operable.

Just as hard on engines as no use is abuse. Hard climbs and fast descents, causing abnormal heating and cooling conditions, are extremely destructive to air-cooled engines. Training aircraft often exhibit this trait due to their usage (takeoff and landing practice).

Beware of the engine that has just a few hours on it since an overhaul. Perhaps something is not right with the overhaul, or it was a very cheap job, just to make the plane more salable.

When it comes to overhauls, I recommend the large shops that specialize in aircraft engine rebuilding. I'm not saying that the local FBO can't do a good job; I just feel that the large organizations specializing in this work have more experience and equipment to work with. In addition, they have reputations to live up to, and most will back you in the event of difficulties.

Here are some typical costs for a complete overhaul (based upon average pricing, including installation):

ENGINE	COST
O-300	$13,000
GO-300	14,000
O-320-E2D	11,000
O-320-H2AD	12,500
O-320-D2J	11,500
O-360	12,500
IO-360	17,000

At the time of this writing (early 1993), AirPower, of Arlington, Texas, was setting a Lycoming factory overhauled O-320-D2J engine for fewer than $9000 and charged $1500 for installation. This is one of the few bargains I have seen in aviation.

AVIATION FUELS

The following information is reprinted by permission of Avco Lycoming, as found in its *Key Reprints*:

We have received many inquiries from the field expressing concern over the limited availability of 80/87 grade fuel, and the associated questions about the use of higher leaded fuel in engines rated for grade 80/87 fuel. The leading fuel suppliers indicate that in some areas 80/87 grade aviation fuel is not available. It is further indicated that the trend is toward phase-out of 80/87 aviation grade fuel. The low lead 100LL avgas, blue color, which is limited to 2 ml tetraethyl lead per gallon, will gradually become the only fuel available for piston engines. Whenever 80/87 is not available, you should use the lowest lead 100-grade fuel available. Automotive fuels should never be used as a substitute for aviation fuel in aircraft engines.

The continuous use, more than 25 percent of the operating time, with the higher leaded fuels in engines certified for 80 octane fuel can result in increased engine deposits both in the combustion chamber and in the engine oil. It may require increased spark plug maintenance and more frequent oil changes. The frequency of spark plug maintenance and oil drain periods will be governed by the amount of lead per gallon and the type of operation. Operation at full rich mixture requires more frequent maintenance periods; therefore, it is important to use properly approved mixture leaning procedures.

To reduce or keep engine deposits at a minimum when using the higher leaded fuels, 100LL avgas blue or 100 green, it is essential that the following four conditions of operation and maintenance are applied [Each condition is subsequently explained]:

- Fuel management required in all modes of flight operation. (*See* A, General Rules.)
- Prior to engine shutdown, run up to 1200 rpm for 1 minute to clean out any unburned fuel after taxiing in. (*See* B, Engine Shutdown.)
- Replace lubricating oil and filters each 50 hours of operation, under normal environmental conditions. (*See* C, Lubrication Recommendations.)
- Proper selection of spark plug types and good maintenance are necessary. (*See* D, Spark Plugs.)

The use of economy cruise engine leaning whenever possible will keep deposits to a minimum. Pertinent portions of the manual leaning procedures as recommended in AVCO Lycoming Service Instruction No. 1094 are reprinted here for reference.

A. General Rules
   1. Never lean the mixture from full rich during takeoff, climb, or high-performance cruise operation unless the airplane owner's manual advises otherwise: however, during takeoff from high elevation airports or during climb at higher altitudes, roughness or reduction of power may occur at full rich mixtures. In such a case, the mixture may be adjusted only enough to obtain smooth engine operation. Careful observation of temperature instruments should be practiced.
   2. Operate the engine at maximum power mixture for performance cruise powers and at best economy mixture for economy cruise power, unless otherwise specified in the airplane owner's manual.
   3. Always return the mixture to full rich before increasing power settings.
   4. During let-down and reduced power flight operations it may be necessary to manually lean or leave the mixture setting at cruise position prior to landing. During the landing sequence, the mixture control should then be placed in the full rich position, unless landing at high elevation fields where leaning may be necessary.
   5. Methods for manually setting maximum power or best economy mixture.
     a. Engine Tachometer—Airspeed Indicator Method: The tachometer

and/or the airspeed indicator may be used to locate, approximately, maximum power and best economy mixture ranges. When a fixed-pitch propeller is used, either or both instruments are useful indicators. If the airplane uses a constant-speed propeller, the airspeed indicator is useful. Regardless of the propeller type, set the controls for the desired cruise power as shown in the owner's manual. Gradually lean the mixture from full rich until either the tachometer or the airspeed indicator are reading peaks. At peak indication, the engine is operating in the maximum power range.

b. For Cruise Power: Where best economy operation is allowed by the manufacturer, the mixture is first leaned from full rich to maximum power, then leaning slowly continued until engine operation becomes rough or until engine power is rapidly diminishing as noted by an undesirable decrease in airspeed. When either condition occurs, enrich the mixture sufficiently to obtain an evenly firing engine or the regain of most of the lost airspeed or engine rpm. Some slight engine power and airspeed must be sacrificed to gain best economy mixture setting.

c. Exhaust Gas Temperature Method (EGT): Refer to Service Instruction No. 1094 for procedure.

Recommended fuel management—Manual leaning will not only result in (fewer) engine deposits and reduced maintenance cost, but will provide (improved) economical operation and fuel saving.

B. Engine Shutdown

The deposit formation rate can be greatly retarded by controlling ground operation to minimize separation of the nonvolatile components of the higher leaded aviation fuels. This rate can be accelerated by (1) low mixture temperatures and (2) excessively rich fuel and air mixtures associated with the idling and taxiing operations; therefore, it is important that engine idling speeds should be set at the proper 600 to 650 rpm range with the idle mixture adjusted properly to provide smooth idling operation. Shutdown procedure recommends setting rpm at 1200 for one minute prior to shutdown.

C. Lubrication Recommendations

Many of the engine deposits formed by the use of the higher leaded fuel are in suspension within the engine oil and are not removed by a full flow filter. When sufficient amounts of these contaminants in the oil reach a high temperature area of the engine they can be baked out, resulting in possible malfunctions such as in exhaust valve guides, causing sticking valves. When using the higher leaded fuels, the recommended oil drain period of 50 hours should not be extended, and if occurrences of valve sticking is noted, all guides should be reamed and (the oil and filter replacement or cleaning period should be shortened).

D. Spark Plugs

Spark plugs should be rotated from the top to bottom on a 50-hour basis, and should be serviced on a 100-hour basis. If excessive spark plug lead fouling occurs, the selection of a hotter plug may be necessary; however, depending on the type of lead deposit formed, a colder plug may better resolve the problem. Depending on the lead content of the fuel and the type of operation, more frequent cleaning of the spark plugs may be necessary. (When) the majority of operation is at low power, such as patrolling, a hotter plug would be advantageous. If the majority of operation is at high cruise power, a colder plug is recommended.

Color coding of avgas

- Red is 80 octane containing 0.50 ml lead per gallon
- Blue is 100 octane containing 2.0 ml lead per gallon
- Green is 100 octane containing 3.0 ml lead per gallon

From the FAA

In April 1977, the use of Tricresyl phosphate (TCP) was approved for use in Lycoming and Continental engines that do not incorporate turbosuperchargers (Fig. 3-3). TCP is a fuel additive that is available from most FBOs and the manufacturer:

Alcor, Inc.
10130 Jones-Maltsberger Rd., P.O. Box 32516
San Antonio, TX 78216
(800) 354-7233
(512) 349-3771

Fig. 3-3. Alcor's TCP fuel treatment.

AUTO FUELS

Considerable controversy and discussion about the use of auto fuels (sometimes referred to as *mogas*) in certified aircraft engines prompted headlines about the issue in most aviation journals. Both sides have pros and cons; however, it is up to the individual pilot to make a choice about the use of nonaviation fuels in an airplane. Consider the following:

Economy. Unleaded auto fuel is certainly less expensive than 100LL. Auto fuel does appear to operate well in the older engines that require 80 octane fuel.

If you have a private gas tank/pump, it might be advantageous to utilize auto fuel. It'll be far easier to locate a jobber willing to keep an auto fuel tank filled than it will be to find an avgas supplier willing to make small deliveries. This is particularly true at private, rural airstrips.

The quality of various gasolines and the additives might be inconsistent. In particular, many low-lead auto fuels have alcohol in them. Alcohol is destructive to certain parts of the typical aircraft fuel system. Engine manufacturers claim the use of auto fuel will void warranty service.

Insurance coverage. Prior to purchasing an auto fuel STC, check with your insurance carrier and obtain approval in writing. Many FBOs are reluctant to make auto fuels available for reasons such as product liability and less profit (Fig. 3-4).

FAA's comment

Here is a partial reprint of Advisory Circular #AC 150/5190-A (April 4, 1972) that addresses restrictions on self-service:

> Any unreasonable restriction imposed on the owners and operators of aircraft regarding the servicing of their own aircraft and equipment may be considered as

Fig. 3-4. This progressive FBO offers options of 100LL avgas and auto fuel to the customer.

a violation of agency policy. The owner of an aircraft should be permitted to fuel, wash, repair, paint, and otherwise take care of his own aircraft, provided there is no attempt to perform such services for others. Restrictions which have the effect of diverting activity of this type to a commercial enterprise amount to an exclusive right contrary to law.

The FAA warns against use of any auto fuel containing ethanol or methanol (IAW AC 23.1521), however, states that auto fuels can lead to reduced maintenance costs (FAA AC 91-33). Information relative to alcohol in fuel can be found in AC 91-40. **Warning:** Never use fuel containing alcohol in an airplane.

If you decide that mogas is to be your fuel of choice, be sure your fuel complies with Specification D-439 and D-4814 by ASTM (American Society for Testing Materials). Generally, compliance should be no problem if you select a major brand. Lesser brands, doing business in the seventeen states not requiring compliance with the above standards, should be checked before use. States not requiring compliance with the ASTM standards:

Alaska	New York
Kentucky	Ohio
Maine	Oregon
Massachusetts	Pennsylvania
Michigan	Texas
Missouri	Vermont
Nebraska	Washington
New Hampshire	West Virginia
New Jersey	

Watch out for progress

For environmental reasons, the Environmental Protection Agency requires certain geographical areas of the United States to use up to 10 percent oxygenates (alcohol or MTBE) in auto fuels, to reduce air pollution. This could have an adverse effect upon the STCs for auto fuels because new auto fuel formulas will not meet the requirements of D-439.

Testing for alcohol in fuel

If you suspect that your choice of fuel might contain alcohol, the following simple test will aid you. The test works because alcohol readily mixes with water, thereby causing it to separate from the fuel and settle to the bottom. A chemist's graduate is recommended for this test due to its easily read scales; get a graduate that is marked in divisions of 10s that holds 10 ounces of liquid, usually available from chemical supply houses and swimming pool supply stores.

1. Mix 9 parts of fuel with 1 part water.
2. Agitate (shake) the mixture.
3. Let stand for 5 minutes.

After the mixture settles out, look for a dividing line between the water and fuel. The water (added as part of the test) and (or) water and alcohol (resulting from the test water mixing with alcohol in the fuel) will settle to the bottom of the test solution. The fuel contains no alcohol if the results show 9 parts fuel and 1 part water. If the fuel tested contains alcohol, the results will show 8 parts fuel and 2 parts of water and alcohol mixture; normally alcohol will make up 10 percent by volume.

Mogas STCs

Petersen Aviation, Inc., has STCs for leaded and unleaded auto fuels. EAA has STCs for only unleaded auto fuels.

Petersen Aviation, Inc.
RR #1 Box 18
Minden, NE 68959
(308) 237-9338

Experimental Aircraft Association
Wittman Field
Oshkosh, WI 54903
(414) 426-4800

COLD WEATHER OPERATION

A well-charged battery plus proper engine lubrication are essential when operating an airplane in cold weather.

Battery power

The aircraft's storage battery provides the necessary cranking power for engine starting. Cold-weather cranking is the heaviest drain possible on a battery; therefore, the battery must be kept in top condition. The condition (charge level) of a storage battery is determined by measuring with a *hydrometer* the specific gravity of the electrolyte fluid in the battery's cells (Fig. 3-5).

Low specific gravity not only indicates that the battery could not meet cranking demands, it also means the battery could freeze in cold weather (Fig. 3-6). A storage battery will be destroyed if it freezes, generally by cracking of the case. If the case cracks, the electrolyte inside the battery (an extremely corrosive acid) might drain into the interior of the airplane.

Prevention is the best medicine. The best thing you can do for a storage battery is to keep it charged at all times. This can be done by:

1. Continuous charging with a *trickle charger* that requires an electric power source.
2. Continuous charging with a solar charging unit that requires no electric power.
3. Remove the battery from the airplane and store it, fully charged, in a warm place.

BATTERY HYDROMETER READINGS

1.280 specific gravity indicates a 100% charged battery

1.250	"	"	"	"	75%	"	"
1.220	"	"	"	"	50%	"	"
1.190	"	"	"	"	25%	"	"
1.160	"	"	"	"	0%	"	"

Fig. 3-5. Battery hydrometer readings.

BATTERY FREEZE TEMPERATURES

1.280 specific gravity indicates a –85° F freeze point

1.250	"	"	"	"	–62	"	"	"	"
1.220	"	"	"	"	–30	"	"	"	"
1.190	"	"	"	"	–11	"	"	"	"

Note: a completely discharged battery will freeze at 19° F.

Fig. 3-6. Battery freeze temperatures.

Engine lubrication

An airplane engine is designed to operate within a particular range of internal temperatures. This temperature results from the heat of fuel combustion minus the ambient air cooling the engine. The oil temperature gauge is the standard indicator for checking that an engine is operating within the desired temperature range. The oil temperature gauge has a green arc on its face, showing the engine's normal operating temperature range. If the gauge indicates the oil has reached a temperature within this range, full power operation is allowed; if the engine runs too hot, it can actually burn up; if the engine runs too cold, it can beat itself to death from poor lubrication or a lack of oil flow, which would be caused by congealed oil.

Engine oil considerations

Check the engine's operating manual to ascertain the correct weight (viscosity) of engine oil for the temperatures you plan to operate in. Oil weight refers to the SAE viscosities as measured in Saybolt Seconds Universal; SSU is the time, in seconds, required for a measured quantity of oil to drip through a small diameter tube at 210°F. Oil designated as W has been tested in a similar manner at 0°F, which means a 20W50 oil drips as if it were a standard 20 weight oil at 0°F and a 50 weight oil at 210°F.

Note that the viscosity numbers are doubled when referring to aviation grade oils: SAE 50 is aviation grade 100. In the past, the general rule of thumb was SAE 50 oil for operation when the ambient air temperature was above 40°F and SAE 30 for operation below 40°F; however, the oil of choice today is a multiweight oil, such as 10W50. The terms multiweight and multiviscosity are interchangeable.

Recall that multiweight oil will flow at low temperatures; a 15W50 multiweight has a pour point of −30°F, yet can be used when ambient air temperatures go as high as 100°F. These wide temperature application ranges are created by combining synthetic oils, and other additives, with straight mineral oils. Although multiweight oils pour at low temperatures, engine preheating is required prior to cold starts because cold oil does not have the lubricating properties of warm oil. The use of multiweight oils reduces the need for seasonal oil changes based upon projected temperature changes; however, this is not a panacea for all cold-weather engine operations.

Multiviscosity oils, designed for operation over a large range of temperatures, do nothing to ease low temperature lubrication problems; however, they do permit easier cranking. Remember, the low temperature of pouring point only means the oil is still a liquid, not a good lubricator.

Cold oil will move inside the engine when forced by the engine oil pump; however, the oil pressure can be limited by the bypass relief valve. The valve protects the oil pump, cooler, and filter from failing because of excessive oil pressure.

As a result of the relief valve action, much of the needed lubricating oil never reaches critical bearing points inside a cold engine. This can become a classical case of the dog chasing its tail: cold oil returns, through the relief valve, to the crankcase; inside the crankcase, the oil never gets a chance to warm to proper lubricating temperatures, and is again pumped out while still cold, unable to lubricate well, and will return still cold.

Warming the engine

Preheat an aircraft engine before attempting a start in cold weather. As a rule of thumb, you will need to preheat when the ambient air temperature falls below 30°F. Check the engine operation manual before any cold starting. Preheating allows the lubricating oil to flow and fuel ignition to take place easier. The strongest case for preheating is reduced wear because the warmer flowing oil lubricates better.

Warming an engine can be accomplished two ways. The first is constant application of heat to the engine. A popular method for providing constant heat is the use of electric heating elements that are semiflexible thin pads permanently affixed to the engine (Fig. 3-7). The second engine warmer is a preheater that is designed to quickly

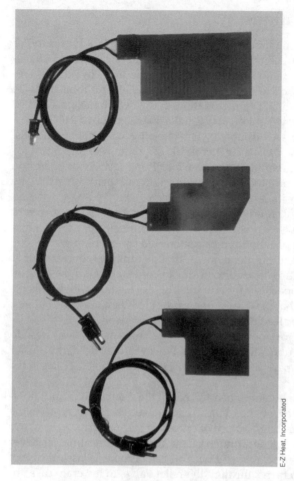

Fig. 3-7. Electric preheaters are easily installed (without STC or form 337), themostatically controlled, and use approximately 300 watts of power per hour.

E-Z Heat, Incorporated

heat up an engine prior to starting (Fig. 3-8). A portable preheater usually has small propane gas cylinders and a forced-air blower.

ENGINE MONITORING

Various gauges and instruments monitor the engine. Remember: Operating temperatures that are too cold or too hot can be dangerous for the engine's integrity and your financial well-being. The tachometer, oil temperature, and oil pressure gauges are all familiar and found on the typical instrument panel; however, other instruments also closely monitor the engine.

Exhaust gas temperature (EGT). The EGT gauge measures the temperature of the exhaust gases as they enter the exhaust manifold. This instrument is extremely valuable for monitoring leaning procedures (Fig. 3-9).

Cylinder head temperature (CHT). The CHT gauge indicates the temperature of the cylinder heads. Problems such as lack of adequate cooling or over cooling (in cold weather) can be detected by it.

Fig. 3-8. The SureStart IV is a compact portable preheater that produces 100,000 BTU.

Fig. 3-9. Digital display of CHT and EGT shows each cylinder by bargraph.

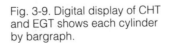

Carburetor ice detector. A carburetor ice detector (Fig. 3-10) detects ice, not just low temperature. Because carb ice is a product of temperature, humidity, and the venturi effect, mere temperature indication is not satisfactory. The detector utilizes an optical probe in the carburetor throat and is so sensitive that it can detect frost up to 5 minutes before ice begins to form, giving the pilot plenty of time to take corrective action. For a complete packet of information about carburetor ice detection contact:

ARP Industries, Inc.
36 Bay Drive East
Huntington, NY 11743
(516) 427-1585

Fig. 3-10. Carburetor ice detector.
ARP Industries, Incorporated

4

Airworthiness directives

AIRPLANES ARE EXTREMELY RELIABLE; unfortunately, from time to time, an airplane requires inspection, service, or repairs as a result of unforeseen problems. A particular problem sometimes affects numerous airplanes of a certain make and model. Required procedures for the inspection, service, repair, or perhaps complete replacement of components to fix the problem are set forth in airworthiness directives. ADs are subject to FAR Part 39 and must be complied with to maintain airworthiness of the airplane. The AD might be a simple one-time inspection, a periodic inspection (perhaps at 25-, 50-, or even 1000-hour intervals), or a major modification to the airframe or engine.

Certain ADs are relatively inexpensive to comply with because they are basically inspections; other ADs can be very expensive, involving extensive engine or airframe modifications and repairs. ADs are not normally handled like automobile recalls with the manufacturer being responsible for the costs involved. Sometimes the manufacturers will offer parts, labor, or both, free of charge, but don't count on it. Even though ADs correct deficient design or poor quality control of parts or workmanship, AD compliance is usually paid for by the owner.

Notice of an AD will be placed in the U.S. government's *Federal Register*, plus the AD is sent by mail to registered owners of the aircraft. In an emergency, the information will be sent by telegram to registered owners. Either way, the purpose is to assure the integrity of the airplane and your safety. The records of AD compliance become a part of the aircraft logbooks. When looking at an airplane to buy, verify AD compliance.

AD LISTS

The following AD list should not be considered the last word; it is only an abbreviated guide to assist the owner or would-be owner or pilot when checking for AD compliance. Listed ADs do not affect every 172 airplane. For a complete check of ADs on an airplane, see your mechanic, or contact the AOPA, which will provide a list of ADs for a particular aircraft (by serial number) for a fee. The highly accurate search is well worth the money spent.

Model 172

59-10-3: Replace/relocate the flasher switch from SN 30000 through 34500 and SN 28000 through 36003.

68-17-4: Test/rework the stall warning system on all models.

68-19-5: Aircraft is removed from the utility category—those 172s with the Franklin GA-335B engine only.

70-10-6: Replace the solid metal oil pressure line with a flexible hose assembly on 172 I and K models.

71-18-1: Install new fuel selector valve placards showing information on 172s SN 17248735 through 17256512.

71-22-2: Inspect/replace the nose gear fork after 1000 hours of operation.

72-3-3: Inspect the flap actuator jack screw each 100 hours of operation.

72-7-2: Install a new fuel selector valve, SN 28000 through 17258855.

73-17-1: Placard the auxiliary fuel pump if so equipped, 172 and 175.

73-23-1: Replace piston pins, Lycoming O-320 engines.

73-23-7: Replace the wing attachment fittings, SN 17261664 through 17261808.

74-4-1: Inspect the aft bulkhead for cracks, SN17260759 through 17261495.

74-6-2: Rework the cabin heater system if Avcon STCed for 180-hp engine.

74-8-1: Inspect the autopilot actuators on all 300/400/800 series units.

74-26-9: Inspect Bendix magnetos for solid steel drive shaft bushing and replace as necessary.

75-8-9: Replace the oil pump shaft and impeller on Lycoming O-320 engines.

76-4-3: Modify the ARC PA-500A actuator gear train on all 300/400/800 series autopilots.

76-21-6: Replace the engine oil cooler on certain early models to prevent oil loss, 172 M and N.

77-2-9: Replace the flap actuator ball nut assembly SN17267789 through 17268239.

77-7-7: Modify the oil dipstick tube to allow accurate monitoring of oil levels on O-320-H2AD Lycoming engines.

77-12-8: Test/modify the ground power and electrical system to prevent undesired turning of the propeller.

77-16-1: Check certain McCauley props (see AD for propeller serial numbers).

77-20-7: Replace the valve tappets and rocker arm studs on O-320-H2AD Lycoming engine.

78-12-8: Replace the oil pump gear on O-320-H2AD Lycoming engine.

78-12-9: Replace the crankshaft on O-320-H2AD Lycoming engine.

79-8-3: Remove/modify the cigarette lighter wiring to prevent possible fire, SN 28000 through 29999, 46001 through 47746: 17247747 through 17250572, and 17259224 through 17267584.

79-10-3: Check the engine mounting bolts on 172 N models.

79-10-14: Install a vented fuel cap and placard various models.

79-13-8: Replace the Airborne dry air pump, if installed after 5-15-79.

79-18-5: Replace LiS02 ELT batteries.

79-18-6: Inspect/modify Bendix magnetos as required per their service bulletin.

80-4-3: Replace the exhaust valve spring seats and hydraulic lifters on O-320-H2AD Lycoming engines to prevent bent push rods.

80-4-8: Install a cover over the map light switch to prevent chafing against the fuel lines and a possible fire on 172 N.

80-6-3: Install a new flap cable clamp 172 M & N models and R172K.

80-6-5: Test Slick magneto couplings on all engines.

80-14-7: Inspect the exhaust valve springs and seats on O-320-H2AD Lycoming engines and replace as needed.

80-17-14: Comply with Bendix magneto service bulletins.

80-25-7: Inspect the Stewart Warner oil cooler; replace as listed in AD.

81-5-1: Inspect the fuel quantity gauges and markings for accuracy on 172 N and 172 RG.

81-15-3: Replace the Brackett engine air filter on all 172s through M model.

81-16-5: Inspect/replace the Slick magneto coil for cracks.

81-16-9: Rework the elevator control system on 172 N models SN 71035 through 74523.

81-18-4: Replace the oil pump impeller and shaft on Lycoming O-320 engines.

82-7-2: Inspect the crankcase breather and seal on 172 through 172 P and 175 models.

82-11-5: Comply with Bendix Service Bulletin 617 on magnetos.

82-13-1: Periodic inspection/replace of the gripper bushing block and check pistons/valves on engines with S-1200 series Bendix magnetos.

82-20-1: Inspect the Bendix impulse couplers prior to 300 hours usage on the magnetos.

83-10-3: Modify the control wheel yoke guide, 172 M SN 66940 through 172 P models.

83-17-6: Rebalance the ailerons on some Robertson STOL conversions.

83-22-6: Inspection of the aileron hinges on 172 N through Q.

84-26-2: Replace the paper air filter elements each 500 hours.

86-24-7: Modify engine control rod ends.

86-26-4: Inspect/modify shoulder harnesses.

87-10-6: Inspect/replace the rocker arm assembly on O-320 (except H).

87-20-3: Inspect/replace seat rails on most models.

90-6-3: Test exhaust heater on all 172s from SN 36216 and up to prevent carbon monoxide from entering the cabin.

91-14-22: Inspect crankshaft gear and counter bolt recess, alignment dowel, etc. at overhaul or after prop strike on most O-320 engines.

Cutlass RG

80-1-6: Modify the flap actuator assembly.

80-6-5: Test the Slick magneto coupling on all engines.

80-19-8: Rework the fuel mixture control assembly on SN 001 through 573.

80-25-7: Inspect the oil cooler and replace as listed in the AD.

81-5-1: Inspect the fuel quantity gauges and markings for accuracy SN 0001 through 0698.

81-14-6: Replace the rudder trim to nose gear bungee, SN 001 through 769.

81-16-5: Inspect/replace the Slick magneto coil for cracks.

81-16-9: Rework the elevator control system on 172 RG models, SN 1 through 789.

81-18-4: Replace the oil pump impeller and shaft on Lycoming O-360 engines.

82-20-1: Inspect the Bendix impulse couplers prior to 300 hours usage on the magnetos.

82-27-2: Dye check the propeller blade shank on certain McCauley propellers.

83-14-4: Modify the cabin heater shroud.

83-22-6: Inspection of the aileron hinges.

84-26-2: Replace the paper air filter elements each 500 hours.

85-20-1: Rework the cabin heater system.

86-19-11: Modify the fuel quick drain system.

86-24-7: Modify engine control rod ends.

86-26-4: Inspect/modify shoulder harnesses.

87-10-6: Inspect/replace the rocker arm assembly on O-360.

87-20-3: Inspect/replace seat rails on most models.

88-3-6: Replace oil filter p/n 649309 & 649310 in IO-360.

91-19-3: Inspect/replace AD specified oil filters on IO-360.

Hawk XP

79-13-8: Replace the Airborne dry air pump if installed after 5-15-79.

79-18-5: Replace LiS02 ELT batteries.

80-6-3: Install a new flap cable clamp.

80-6-5: Test Slick the magneto coupling on all engines.

81-5-1: Inspect the fuel quantity gauges and markings for accuracy.

81-13-10: Rework the oil pump drive on certain IO-360 engines.

81-16-5: Inspect/replace the Slick magneto coil for cracks.

81-16-9: Rework the elevator control system on R172K models, SN R1722930 through R1723425.

82-20-1: Inspect/service Bendix magneto.

82-27-2: Dye check the propeller blade shank on certain McCauley propellers.

83-10-3: Modify the control wheel, R172 K models.

83-22-6: Inspection of the aileron hinges.

84-26-2: Replace the paper filters each 500 hours.

87-20-3: Inspect/replace seat rails on most models.

88-3-6: Replace oil filter p/n 649309 & 649310 in IO-360.

91-19-3: Inspect/replace AD specified oil filters on IO-360.

92-4-9: Rework cylinder assembly per service bulletin or retorque rocker arm assembly if assembly complies.

Model 175

62-22-1: Reinstall the vacuum pump on Continental O-300 A engines.

63-22-3: Rework the carburetor on all models.

71-22-2: Inspect/replace the nose gear fork after 1000 hours of operation.

73-17-1: Placard the auxiliary fuel pump if so equipped.

74-26-9: Inspect Bendix magneto for solid steel drive shank bushing. Replace as necessary.

77-16-1: Check certain McCauley props (see AD for propeller serial numbers).

79-8-3: Remove/modify the cigarette lighter wiring.

79-10-14: Install a vented fuel cap and placard same.

79-18-5: Replace LiS02 ELT batteries.

82-7-2: Inspect the crankcase breather and seal on 175 models.

82-20-1: Inspect the Bendix impulse couplers prior to 300 hours usage on the magnetos.

84-26-2: Replace air filter each 500 hrs.

86-26-4: Inspect/modify shoulder harness.

87-20-3: Inspect/replace seat rails on most models.

DANGEROUS SEATS

The *1968 Cessna Service Manual* for 172 series airplanes makes a point: "WARNING: It is extremely important that the pilot's seat stops are installed, since acceleration and deceleration could possibly permit the seat to become disengaged from the seat rails and create a hazardous situation, especially during takeoff and landing." This is very serious. How would you feel if you had just started a climbout and suddenly were pitched over backward, seat and all? Regaining control of the aircraft would be impossible.

The pilot's seat is mounted to two aluminum tracks (rails) and the seat slides back and forth for adjustment. A pin holds the seat position on the track.

The following is quoted from *Airworthiness Alerts*, a monthly publication from the FAA Aviation Standards National Field Office in Oklahoma City:

Cessna Single Engine Models

Numerous reports indicate that difficulties continue to be encountered with seat attachments, structures, locking mechanisms, tracks, and stops. When required inspections are made, it is suggested the following items be examined:

1. Check the seat assembly for structural integrity.
2. Inspect the roller brackets for separation and wear.
3. Examine the locking mechanism (actuating arm, linkage, locking pin) for wear and evidence of impending failure.
4. Inspect the floor-mounted seat rails for condition and security, locking pin holes for wear, and rail stops for security.
5. Determine that the floor structure in the vicinity of the rails is not cracked or distorted.

Defective or worn parts are a potential hazard which should be given prompt attention. Accomplish repair and/or replacement of damaged components in accordance with the manufacturer's service publications.

Note: This article was previously published in Alerts No. 32, dated March 1981. The same type problems are still being reported. The NTSB (National Transportation Safety Board) has identified these same seat problems as the probable cause in several fatal accidents.

Always keep this seat problem in mind at preflight time, and always check the seats after locking them in place. AD 87-20-3 now applies to most Cessna single engine airplanes and requires an inspection of the seat rails and replacement if required. Replacement seat rails are available from: McFarlane Aviation, Inc., RR #3, Baldwin, KS 66006, (800) 544-8594.

H2AD ENGINE

The Avco Lycoming O-320-H2AD engine was designed to burn 100LL avgas; it produced 160 horsepower and was built at Cessna's request for use in the Model 172s. When the engine debuted in 1977, 100LL was practically the only fuel available for piston-powered airplanes. The H2AD was less complicated internally than its predecessors. It was designed to require only a minimum amount of machining after casting.

The H2AD engine was supposed to be cheaper to build than other aircraft engines, but proved to be anything but economical.

From 1977 until 1981, when the H engine was replaced with the D2J engine, Cessna 172 owners were plagued with expensive engine problems. The major problem was disintegration (*spalling*) of the valve tappet and cam surfaces where they make required mechanical contact. The small metal particles then were scattered in the engine lubrication system, quickly ruining the engine.

Lycoming redesigned the tappets, camshaft, and rocker arm studs but that did not help because H engines were still dropping dead. Cessna's 1978 Blue Streak program was supposed to completely alleviate all the past problems of the H2AD engine. That did not work. Mandatory oil additives and special operating procedures had no effect because the spalling continued in the H2AD engines.

It's interesting to note that Cessna has attempted to alleviate the problem through various changes and modifications. It even offered a 2000-hour engine warranty, on a prorated basis, to no avail. Perhaps the final fix for the H2AD was made from serial number 7976 and up. A heavier crankcase was redesigned to allow the use of a larger hydraulic tappet.

The end result of all the problems with the H2AD engine was a lot of lost flying time, loads of owner money spent on repairs, and plenty of frustration for all parties concerned.

The Aviation Consumer reported in November 1983 that a California court had ordered Cessna to pay damages of $25,000 to a hapless purchaser of a 172 with an H engine. This particular owner's successful suit was based upon a fraud statute: fraud by misrepresentation. The plaintiff felt the aircraft was misrepresented to him due to the many problems he encountered with the O-320-H2AD engine. It was misrepresented in that the plane was not safe, reliable, or of use as advertised. The court agreed.

It is very common for engine rebuilders to charge lower prices for work on the earlier E model or the later D version of the O-320 engine. Generally, the H2AD will cost approximately 10 percent more.

On the brighter side, some 172 owners like their planes so much that they have replaced the H engine with a 180-hp engine under a supplemental type certificate. For further information on STCs, *see* chapter 10.

Finally, the H2AD problems are unique to the 172 because the engine was never installed on any other production aircraft.

CESSNA SERVICE BULLETINS

Cessna offers a service bulletin listing program as a paid service for owners of Cessna airplanes. Cessna claims that the service "will keep you informed regarding all service bulletins and service information letters issued by Cessna." The listing service only discloses which bulletins apply to a specific airplane. The actual bulletins or service letters must be purchased separately; cost is nominal. For more information about this service, contact:

Cessna Aircraft Company, Support Services Department
P.O. Box 7704
Wichita, KS 67277
(316) 946-6118
FAX (316) 942-9006

5

Selecting a 172

THE SEARCH FOR A GOOD USED CESSNA 172 does not usually have to be wide and exhaustive because so many have been built. This is good for the buyer because an adequate selection to choose from is assured. Before going into the marketplace and searching for an airplane for purchase, prepare yourself with as much background information as possible. Read old magazine articles about the 172 series and talk to present and past owners. Ask questions and listen carefully to the answers. Beware the individual bearing sour grapes; sort the wheat from the chaff; be sure of your sources because not all speak the truth, whether innocently or otherwise. The good and reliable information shall prove to be very beneficial in the long run.

THE SEARCH

Because many models and price ranges are available, the purchaser is encouraged to set a range of personal expectations. This could be based upon features desired, options available, or—most likely—the cash available for such a purchase.

Usually, one starts looking at home. If you know and are professionally comfortable with an FBO, then perhaps this is a good way to search. Explain what you are looking for. Often an FBO will know of airplanes that are for sale or will come on the market soon; the FBO is an insider to the business and might know the owner of a 172 who is shopping for a bigger airplane, but has not started to advertise the 172 for sale.

If there is nothing of interest at your airport, then broaden the search. Check the bulletin boards at other local airports. Ask around while you're there, then walk around and look for airplanes with "For Sale" signs in the windows. You could even put an "Airplane Wanted" ad on the bulletin boards. The local newspaper will sometimes have airplanes listed in the classified ads, especially in the Sunday edition of a large metropolitan newspaper.

Several aviation publications have become leaders in airplane advertising. Among them are *Trade-A-Plane* and *General Aviation News and Flyer*.

If you are really serious and want to find that illusive bargain you need to be at the front of the line. This means getting *Trade-A-Plane* before everyone else does. Subscriptions are available as first class U.S. mail, or better yet, as Federal Express 2nd day air priority delivery.

General Aviation News and Flyer is a twice monthly newspaper carrying up-to-the-minute news and information concerning general aviation. "You read it here months before the magazines print it" is a phrase I have heard more than one pilot say about *General Aviation News and Flyer*. Their classifieds are printed on pink paper and carried as an insert to the newspaper and will typically number over twenty pages.

These two and the other publications listed might be nice to have around whether or not you are shopping:

Trade-A-Plane
P.O. Box 509
Crossville, TN 38555
(615) 484-5137

General Aviation News and Flyer
P.O. Box 98786
Tacoma, WA 98498-0786
(206) 588-1743

Atlantic Flyer
Civil Air Terminal, Hanscom Field
Bedford, MA 01730
(617) 274-7208

Aviators Hot Line
1003 Central Ave.
Fort Dodge, IA 50501
(515) 955-1600

A/C Flyer
P.O. Box 609
Hightstown, NJ 08520
(609) 426-7070

In Flight
P.O. Box 620477
Woodside, CA 94062
(415) 364-8110

A dealer in Texas handles a large volume of Cessna airplanes:

Van Bortel Aircraft, Inc.
4900 South Collins, Arlington Municipal Airport
Arlington, TX 76018
(800) 359-4295
(817) 468-7788

Listings of used airplanes can be found in magazines (*AOPA Pilot, Flying, Plane & Pilot, Private Pilot*), but remember that the ads are stale. Magazines typically have a 60–90-day lag time between ad placement and printing.

Classified advertisements

Most airplane advertisements make use of abbreviations that describe the individual airplane and its equipment. Here's a sample ad:

62 C172, 3103TT, 435 SMOH, May ANN, FGP,
Dual NAV/COM, GS, MB, ELT, NDH. $18,750 firm.
999-555-1234

Translated: A 1962 Cessna 172 airplane with 3103 total hours on the airframe and an engine with 435 hours since a major overhaul. The next annual inspection is due in May. It is equipped with a full gyro instrument panel, has two navigation and communication radios, a glideslope receiver and indicator, a marker beacon receiver, an emergency locator transmitter, and, best of all, the airplane has no damage history. The price is $18,750, and the seller will not bargain; however, most sellers do bargain. (A lot of information was crammed inside three lines.)

The ads will have a telephone number, but seldom mention where the airplane is located; the area code reveals the general geographical area—would you want to buy a 172 that has been based for five years in a saltwater climate (subjected to corrosives) or in a desert climate (fewer corrosives, if any)?

Appendix A lists popular advertising abbreviations and appendix B lists area code locations.

Use the telephone

Call the phone number listed in the ad, ask questions, and take very good notes to refer to later. If necessary, ask the seller to stop or speak slower while you are taking notes. While you are asking questions, remember that the telephone inquiry is to determine if you would like to see the airplane, or eliminate it from further consideration; take notes with this in mind.

What is the general appearance and condition of the plane?
How many total hours on the airframe?
How many hours on the engine since new?
How many hours since the last overhaul?
What type of overhaul was done?
Who did the overhaul?
Is there any damage history?
What is the asking price?

Traveling around

Searching for a good used airplane can be expensive and very time-consuming. You'll read ads, make long-distance telephone calls and even travel a great distance to see the airplanes that are advertised.

Sometimes, when an advertisement has appealed to me, I have called the seller, listened to the spiel about the wonderful airplane, then spent my time and my money to travel halfway across the country to look at the airplane. All too often the airplane that was represented on the telephone as a 10 turned out to be a gasping 2 or 3. Perhaps in the eyes of the owner the airplane was a 10, but not in mine. I would never purchase an airplane unseen, but it is done every day and very successfully by many people. The choice, and risk, is up to the purchaser.

Some of these sight-unseen sales are handled by merely reading the ad and conversing with the seller; other sales involve photographs and copies of the logbooks; perhaps a seller will lend a videotape of the airplane to the seller for replay in a VCR.

No bargains

Few bargains exist in the world and used airplanes are not an exception to the rule. An airplane selling for a price that is lower than it should be, probably has a reason for being so low. Even if you know the reason and think it is not as bad as it sounds, get professional advice before buying a bargain airplane because the bargain might become a bust.

Even simple repairs to an airplane call for a common prime ingredient—money, lots of it. For example, replacement of a typical four-place airplane's interior can cost more than $3000, paint jobs cost in excess of $3500, and to refurbish and update a panel with instruments and avionics can cost more than the airplane. Although you might be able to save part of the expenses by doing the work yourself, under competent supervision, small jobs have a way of turning into major time eaters.

Never consider a "back-of-the-hangar" airplane, which is an airplane that does not have an engine, or has long been out of service, or "only needs a few parts" to get it going again; it does not represent economical airplane ownership.

Dollarwise, it is good to consider the initial purchase price of an airplane as the tip of the iceberg. You can either pay a higher price for a plane in top shape, or pay less for a plane requiring work and additional money for maintenance. In the end, the total spent will be nearly equal.

Damaged airplanes

No damage history was mentioned in the sample ad. Although it is generally understood that airplanes with prior damage are worth less in the market than those pristine unblemished examples, a properly repaired airplane should not be shied away from. Properly repaired is the key to this statement. To determine whether the airplane in question was properly repaired requires a competent mechanic to check the logs, work orders for the repairs, and the actual work done.

If the repairs were done at an FAA certified repair station you can normally assume the work was done to very high standards and that the damaged airplane has been restored to original specifications. If the repairs were made by a shade-tree operation, even when inspected by a mechanic with an inspection authorization, exercise due caution before committing yourself to purchase. A basic rule: If any repairs are discernible to the eye, stay away from the airplane.

PREPURCHASE INSPECTION

The object of the prepurchase inspection of a used airplane is to preclude the purchase of a dog. No one wants to buy someone else's troubles. The prepurchase inspection must be completed in an orderly, well-planned manner. Take your time during this inspection; a few extra minutes now could well save you thousands of dollars later—perhaps immediately if you decide not to buy the airplane. Buyer beware: It is your money and your safety.

Why are you selling?

The first inspection item is the most-asked question of anyone selling anything: "Why are you selling it?" Of course, if the seller has something to hide, you can't believe the answer. Fortunately, most people will answer honestly. Often the owner is moving up to a larger airplane and, if so, will start to tell you all about the prospective purchase.

Let the seller talk; listen to determine the owner's flying habits and how the airplane was treated. Perhaps other commitments—spouse says sell or the expenses cannot be met—are forcing the matter. An owner anxious to sell might put you in an advantageous situation, but if the owner is having financial difficulties, consider the quality of maintenance that was performed on the airplane—just enough to pass an inspection or well enough to last several hundred hours. Be sure to ask the seller about any known problems or defects with the airplane. Again, the seller will probably be honest.

DEFINITIONS

Airworthy. The airplane must conform to the original type certificate, or those supplemental type certificates issued for this particular airplane (by serial number). In addition, the airplane must be in safe operating condition relative to wear and age.

Annual inspection. All small airplanes must be inspected annually by either an FAA-certified airframe and powerplant mechanic who holds an inspection authorization, or an FAA-certified repair station, or the airplane's manufacturer. This is a complete inspection of the airframe, powerplant, and all subassemblies.

100-hour inspection. Is of the same scope as the annual, and is required on all commercially operated small airplanes (rental, training, charter, and the like), and must be accomplished every 100 hours of operation. This inspection may be performed by an FAA-certified airframe and powerplant mechanic. An annual inspection will fulfill the 100-hour requirement, but the reverse is not true.

Preflight inspection. A thorough visual inspection of an aircraft prior to flight to spot any obvious problems with the exterior, interior, and engine of the airplane.

Preventive maintenance. FAR Part 43 lists a number of maintenance operations that are considered preventive in nature and may be performed by the owner and pilot of an airplane not flown for commercial reasons. Preventive maintenance is detailed in chapter 7.

Repairs and alterations. Repairs and alterations are major or minor. Major repairs and alterations must be approved—for a return to service—by an FAA-certified airframe and powerplant mechanic holding an inspection authorization, a repair station, or by the FAA. Minor repairs and alterations may be returned to service by an FAA-certificated airframe and powerplant mechanic.

Airworthiness directives. The ADs must be complied with (*see* chapter 4) because they are a required maintenance and repair procedure. Files of ADs and their requirements are kept by mechanics and the FAA flight standards district offices. AD compliance verification is part of the annual inspection.

Service difficulty reports. Difficulty reports are prepared by the FAA from malfunction or defect reports from operators and mechanics. SDRs are not law as an AD; however, they should be adhered to for your own safety.

PERFORMING A PREPURCHASE INSPECTION

The prepurchase inspection has three parts: walk-around inspection and logbook check, test flight, and mechanic's inspection. An inspection determines the worth of the airplane plus the effects of wear and tear, generally caused by five basic sources:

- Weather in the form of exposure to moisture causing oxidation, corrosion, and finish deterioration.
- Friction between moving parts causing wear, erosion, galling, and other physical damage.
- Overload stresses to the airframe from excessive loads during hard landings, overweight operations, and exposure to high winds.
- Heat from the sun or from the engine. Engine heat might be direct (from the exhaust system) or indirect (radiated from the source, such as the engine or engine parts).
- Vibration that causes metal fatigue, structural failure, and excessive structure clearances.

THE WALK-AROUND

The walk-around is really a very thorough preflight divided into four simple and logical steps.

Cabin

Open the cabin door and look inside, notice the general condition of the interior. Care given to the interior can be a good indication of what care was given to the remainder of the airplane. Does the interior appear clean, or recently scrubbed after a long period of inattention? Look in the corners. Does the air smell musty and damp? Is the headliner in one tight piece, and the upholstery unfrayed? What is the condition of the door panels?

Look at the instrument panel. Are the instruments in good condition or are knobs missing and glass faces broken? Is the equipment original or upgraded? Are all upgraded instruments clean and workable? Too often, upgrading—particularly in avionics—is haphazard, with results that are neither attractive nor workable.

Look out the windows. Are they clear, not yellowed, and uncrazed? Side windows are comparatively inexpensive and can be replaced by a do-it-yourselfer. A windshield is expensive and proper installation by a professional is critical. Check the operation of the doors, which should close and lock with little effort. No outside light should be seen around edges of the doors. Check the seats for freedom of movement and adjustability. Check the seat tracks and the adjustment locks for damage. The seat tracks and locks are a problem on most Cessna airplanes.

Airframe

Do a complete walk-around of the airplane. Is the paint in good condition? A paint job is expensive—more than $3500—yet necessary for the protection of the metal surfaces from corrosive elements. The paint job should also please the eye of the beholder.

Dents, wrinkles, or tears of the metal skin might indicate prior damage or careless handling. Each discrepancy must be examined very carefully by an experienced mechanic. Total consideration of all the dings and dents should tell if the airplane has had an easy or a rough life.

Corrosion or rust on skin surfaces or control systems should be cause for alarm. Corrosion is to aluminum what rust is to iron: destructive. Any corrosion or rust should be brought to the attention of a mechanic for review. Corrosion that appears as only minor skin damage might continue, unseen, into the interior structure. Damage such as this creates dangerous structural problems that can be very costly to repair. A "little" corrosion in the wing structure is similar to having a "little" lung cancer—"little" is irrelevant.

Check for fuel leaks around the wings, in particular where the wings attach to the fuselage. If leakage evidence is seen, have a mechanic check the source.

The landing gear should be checked for evidence of being sprung. Check the tires for signs of unusual wear that might indicate other structural damage. Be sure the tires are in good condition. Look at the nosewheel oleo strut for signs of fluid leakage and proper extension.

Move all the control surfaces to assure free and smooth movement. Check each surface for damage. When the controls are centered, the surfaces should also be centered; if they are not centered, a problem in the control rigging exists.

Engine

Open or remove the cowling to inspect the engine and engine area. Search for signs of oil leakage by looking at the engine, the inside of the cowl, and on the firewall. If the leaks are bad enough, oil will be dripping to the ground or onto the nosewheel. Naturally, the seller has probably cleaned all the old oil drips away; however, oil leaves stains. Look for these stains.

Check all the fuel and vacuum hoses and lines for signs of deterioration or chafing. Also check the connections for tightness or signs of leakage. Check engine control linkages and cables for obvious damage and ease of movement. Be sure none of the cables are frayed. Check the battery cables, battery box, and battery for corrosion or other damage.

Check the propeller for damage, such as nicks, cracks, or gouges. Even very small defects can cause stress areas on the prop (*see* chapter 6). Any visible damage to a propeller must be checked by a mechanic. Also check it for movement that would indicate propeller looseness at the hub.

Check the exhaust pipes for rigidity, then reach inside them by rubbing a finger along the inside wall. If your finger comes back perfectly clean, you can be assured that someone has cleaned the inside of the pipe—possibly to remove the oily deposits that form there when an engine is burning a lot of oil. If your finger comes out of the pipe covered with a black oily goo, have your mechanic determine the cause. It could only be a carburetor in need of adjustment. It could also be caused by a large amount of oil blow-by, the latter indicating that the engine needs an expensive overhaul. A light gray dusty coating indicates proper operation.

Check for exhaust stains on the belly of the plane to the rear of the exhaust pipe. This area has probably been washed, but look anyway. If you find black oily goo, then, as above, see your mechanic.

Logbooks

If you are satisfied with what you've seen up to this point, then go back to the cabin, have a seat, and check that all required paperwork is with the airplane. These items are required by the FARs to be in the plane when flown (except for the logs, which must be available):

- Airworthiness certificate
- Aircraft registration certificate
- FCC station license
- Flight manual or operating limitations
- Logbooks (airframe, engine and propeller)
- Current equipment list
- Weight and balance data and chart

Pull out the logbooks and start reading them. Sitting there will also allow you to look once again around the cockpit.

Be sure you're looking at the proper logs for this particular aircraft, and that they are the original logs. Sometimes logbooks get lost and are replaced with new ones. This can happen because of carelessness or theft. This is why many owners do not keep their logs in the plane, and may only provide copies for a sales inspection. Replacement logs might be lacking very important information, or could be outright frauds. Fraud is not unheard of in the used airplane business. Be on your guard if the original logs are not available.

Start with the airframe log by looking in the back for the AD compliance section (chapter 4 has a basic list of Cessna 172 ADs). Check that the list is up-to-date, and that any required periodic inspections have been made. Now go back to the most recent entry; it probably is an annual or 100-hour inspection. The annual inspection will be a statement that reads:

March 21, 1985 Total Time: 3126 hrs.
I certify that this aircraft has been inspected in accordance with an annual inspection and was determined to be in airworthy condition.
Signature
IA # 0000000

From this point backward toward the first entry in the logbook, you'll be looking for similar entries, always keeping track of the total time for continuity purposes and to reveal the number of hours flown between inspections. Also, you will be looking for indications of major repairs and modifications that will be signaled by the phrase, Form 337 filed. A copy of this form should be with the logs, and will tell what work was done. The work might also be described in the logbook.

Form 337, Major Repair and Alteration, is filed with the FAA, and copies are a part of the official file of each airplane. The file, with any 337s, is retrievable from the FAA for a fee.

The engine log will be quite similar in nature to the airframe log, and will contain information from the annual or 100-hour inspections. Total engine time will be given, and possibly an indication of time since any overhaul work, although you might have to do some math here. It's quite possible that this log and engine will not be the originals for the aircraft. As long as the facts are well-documented in both logs, there is no cause for alarm. After all, this would be the case if the original engine was replaced with a factory rebuilt one, or even a used engine from another plane.

Pay particular attention to the numbers that indicate the results of a differential compression check. These numbers are the best single indicator of the overall health of an engine.

Each number is given as a fraction, with the bottom number always being 80. The 80 indicates the air pressure that was utilized for the check; 80 pounds per square inch (psi) is the industry standard. The top number is the air pressure that the combustion chamber was able to maintain while being tested. A perfect 80 psi is not attainable; therefore, the result will always be fewer than 80 psi. The reason for the lower number is the air pressure loss that results from loose, worn, or broken rings; scored or cracked cylinder walls; or burned, stuck, or poorly seated valves. Mechanics can determine which of the above is the cause and, of course, repair the damage.

Normal readings would be no less than 70/80, and should be uniform (within 2 or 3 lbs) for all cylinders. A discrepancy between cylinders could indicate the need of a top overhaul of one or more cylinders. The FAA says that a loss in excess of 25 percent is cause for further investigation. That would be a reading of 60/80, which indicates a very tired engine in need of considerable work and expenditures.

Read the information from the last oil change to determine if debris was found on the oil screen or in the oil filter. Realize that oil changes are often performed by owners, and might not be recorded in the log, even though FARs require all maintenance to be logged. If the oil changes are recorded, how regular were they? I prefer every 25 hours, but 50 is the norm. If a record of oil analysis is available, ask for that information.

If the engine has been top overhauled or majored, the work performed will be described, including a date and the total time on the engine when the work was accomplished.

Check to see if the ADs have been complied with, and the appropriate entries made in the log (*see* the list of basic ADs in chapter 4). **Note:** If an AVCO Lycoming O-320-H2AD is installed on the airplane, be very particular about AD compliance—and listen to your mechanic's advice before purchase (*see* chapters 3, 4, and 12).

THE TEST FLIGHT

The test flight determines if the airplane feels right to you. The flight should last from 30 minutes to two hours. For insurance purposes, I recommend that either the owner or a competent flight instructor approved by the owner accompany you on the test flight. This will also eliminate problems of currency, ratings, and the like, with the FAA, and it will foster better relations with the owner.

After starting the engine, pay particular attention to the gauges. Do they jump to life, or are they sluggish? Watch the oil pressure gauge in particular. Did the oil pressure rise within a few seconds of start? Check the other gauges for proper readings from start-up throughout the flight: ground run-up, takeoff, climbout, cruise, maneuvering, descent, landing, taxi, and shutdown. Do the numbers match those called for in the operations manual? In order to pay more attention to the gauges it might be advisable to have the other pilot make the takeoff.

After you're airborne, check the gyro instruments. Be sure they are stable. Check the ventilation and heating system for proper operation. Do a few turns, stalls, and some level flight. Does the airplane perform as expected? Can it be trimmed for hands-off flight straight-and-level?

Check all the avionics for proper operation. A complete check might require a short cross-country flight to an ILS-equipped airport. That's all right; use the time for additional familiarization. When landing or shooting touch-and-goes, beware of nose-wheel shimmy.

Upon engine shutdown on the ramp, open the engine compartment and look again for oil leaks. Also, check along the belly for indications of oil leakage and blowby. A short flight should be enough to dirty things up again, if they had been dirty to begin with.

If, after the test flight, you decide not to purchase the airplane, it would be ethical to at least offer to pay for the fuel used.

MECHANIC'S INSPECTION

If you are still satisfied with the airplane and desire to pursue the matter further, then have it inspected by an A&P or AI. This inspection might cost a few dollars; however, it could save you thousands. The average for a prepurchase inspection is three to four hours labor, at shop rates. That might be fewer than $100 at a small FBO or several hundred dollars at a large operation.

The mechanic will accomplish a search of ADs, a complete check of the logs, and an overall check of the plane. A compression check and a *borescope* examination must be made to determine the internal condition of the engine. A borescope examination means looking into a cylinder and viewing the top of the piston, the valves, and the cylinder walls with the borescope device.

The mechanic should also check the avionics to determine if they work and are installed in an approved manner and properly noted on the weight and balance sheet.

ADVICE

Always use your own mechanic for the prepurchase inspection: someone you are paying to watch out for your interests, not someone who might have an interest in the sale of the plane, such as an employee of the seller. Have the plane checked even if an annual was just done, unless you know and trust the IA who did the inspection. You might be able to make a deal with the owner over the cost of the mechanic's inspection, particularly if an annual is due.

It's not uncommon to see airplanes listed for sale with the phrase annual at date of sale. I am always leery of this because I don't know who did the annual, or how complete the annual was. All annuals are not created equal. An annual at the date of sale is coming with the airplane, done by the seller, as part of the sale. Who is looking out for your interests?

My standard advice is that if an airplane seller refuses you anything that has been mentioned in this chapter, say thanks, walk away, and look elsewhere. Do not let a seller control the situation. Your money, your safety, and possibly your very life are at stake. Airplanes are not hot sellers, and there is rarely a line forming to make a purchase. You are the buyer; you have the final word.

OWNERSHIP PAPERWORK

Assuming that you have completely inspected your prospective purchase, and found it acceptable at an agreeable price, you're ready to complete the paperwork that will lead to ownership.

Title search

The first step is to assure the airplane has a clear title. A title search is accomplished by checking the aircraft records at the Mike Monroney Aeronautical Center in Oklahoma City, Oklahoma. These records include title information, chain of ownership, major repair or alteration (Form 337) information, and other data pertinent to a particular airplane. The FAA files this information by N-number.

The primary object of a title search is to ensure no liens or other hidden encumbrances are on file against the ownership of the airplane. This search may be done by you, your attorney, or other representative selected by you. AOPA offers inexpensive title insurance, which protects the owner against unrecorded liens, FAA recording mistakes, or other clouds on the title.

Because most prospective purchasers would find it inconvenient to travel to Oklahoma City to do the search themselves, it is advisable to contract with a third party specializing in this service to do the searching. Among other similar service companies (that advertise in *Trade-A-Plane*) are:

King Aircraft Title, Inc.
1411 Classen Blvd. Suite 114
Oklahoma City, OK 73106
(800) 688-1832

AOPA Aircraft and Airmen Records Dept.
P.O. Box 19244, Southwest Station
Oklahoma City, OK 73144
(800) 654-4700

Documents

The following documents must accompany the airplane:

- Bill of sale
- Airworthiness certificate
- Airframe logbook
- Engine and propeller logbook

- Equipment list (including weight and balance data)
- Flight manual

Forms to be completed

Changing official ownership of an airplane requires completion of several FAA and FCC forms. AC Form 8050-2 (Figs. 5-1 and 5-2), Bill of Sale, is the standard means of recording transfer of ownership.

UNITED STATES OF AMERICA

DEPARTMENT OF TRANSPORTATION — FEDERAL AVIATION ADMINISTRATION

AIRCRAFT BILL OF SALE INFORMATION

PREPARATION: Prepare this form in duplicate. Except for signatures, all data should be typewritten or printed. *Signatures must be in ink.* The name of the purchaser must be identical to the name of the applicant shown on the application for aircraft registration

When a trade name is shown as the purchaser or seller, the name of the individual owner or co-owners must be shown along with the trade name.

If the aircraft was not purchased from the last registered owner, conveyances must be submitted completing the chain of ownership from the last registered owner, through all intervening owners, to the applicant.

REGISTRATION AND RECORDING FEES: The fee for issuing a certificate of aircraft registration is $5.00. An additional fee of $5.00 is required when a conditional sales contract is submitted in lieu of bill of sale as evidence of ownership along with the application for aircraft registration ($5.00 for the issuance of the certificate, and $5.00 for recording the lien evidenced by the contract). The fee for recording a conveyance is $5.00 for each aircraft listed thereon. (There is no fee for issuing a certificate of aircraft registration to a governmental unit or for recording a bill of sale that accompanies an application for aircraft registration and the proper registration fee.)

MAILING INSTRUCTIONS:

If this form is used, please mail the original or copy which has been signed in ink to the FAA Aircraft Registry, P.O. Box 25504, Oklahoma City, Oklahoma 73125.

Fig. 5-1. 8050-2, Aircraft Bill of Sale Information.

AC Form 8050-1 (Figs. 5-3 and 5-4), Aircraft Registration, is filed with the Bill of Sale, or its equivalent. If you are purchasing the airplane under a contract of conditional sale, then that contract must accompany the registration application in lieu of the AC Form 8050-2. The pink copy of the registration is retained by you, and will remain in the airplane until the new registration is issued by the FAA.

AC 8050-41 (Fig. 5-5), Release of Lien, must be filed by the seller if a lien is recorded.

AC 8050-64, Assignment of Special Registration Number, is for vanity registration numbers. All U.S. aircraft registration numbers consist of the prefix N, and are followed by: one to five numbers, or one to four numbers and a single-letter suffix, or one to three numbers and a two-letter suffix.

FORM APPROVED
OMB NO. 2120-0042

UNITED STATES OF AMERICA
DEPARTMENT OF TRANSPORTATION FEDERAL AVIATION ADMINISTRATION

AIRCRAFT BILL OF SALE

FOR AND IN CONSIDERATION OF $ THE UNDERSIGNED OWNER(S) OF THE FULL LEGAL AND BENEFICIAL TITLE OF THE AIRCRAFT DESCRIBED AS FOLLOWS:

UNITED STATES
REGISTRATION NUMBER **N**

AIRCRAFT MANUFACTURER & MODEL

AIRCRAFT SERIAL No.

DOES THIS   DAY OF   19

HEREBY SELL, GRANT, TRANSFER AND
DELIVER ALL RIGHTS, TITLE, AND INTERESTS
IN AND TO SUCH AIRCRAFT UNTO:

Do Not Write In This Block
FOR FAA USE ONLY

PURCHASER

NAME AND ADDRESS
(IF INDIVIDUAL(S), GIVE LAST NAME, FIRST NAME, AND MIDDLE INITIAL.)

DEALER CERTIFICATE NUMBER

AND TO   EXECUTORS, ADMINISTRATORS, AND ASSIGNS TO HAVE AND TO HOLD
SINGULARLY THE SAID AIRCRAFT FOREVER, AND WARRANTS THE TITLE THEREOF.

IN TESTIMONY WHEREOF   HAVE SET   HAND AND SEAL THIS   DAY OF   19

SELLER

NAME (S) OF SELLER (TYPED OR PRINTED)	SIGNATURE (S) (IN INK) (IF EXECUTED FOR CO-OWNERSHIP, ALL MUST SIGN.)	TITLE (TYPED OR PRINTED)

ACKNOWLEDGMENT   (NOT REQUIRED FOR PURPOSES OF FAA RECORDING; HOWEVER, MAY BE REQUIRED BY LOCAL LAW FOR VALIDITY OF THE INSTRUMENT.)

ORIGINAL: TO FAA

AC FORM 8050-2 (8-85) (0052-00-629-0002)

Fig. 5-2. 8050-2, Aircraft Bill of Sale.

UNITED STATES OF AMERICA-DEPARTMENT OF TRANSPORTATION

FEDERAL AVIATION ADMINISTRATION-MIKE MONRONEY AERONAUTICAL CENTER

AIRCRAFT REGISTRATION INFORMATION

PREPARATION: Prepare this form in triplicate. Except for signatures, all data should be typewritten or printed. Signatures must be in ink. The name of the applicant should be identical to the name of the purchaser shown on the applicant's evidence of ownership.

EVIDENCE OF OWNERSHIP: The applicant for registration of an aircraft must submit evidence of ownership that meets the requirements prescribed in Part 47 of the Federal Aviation Regulations. AC Form 8050-2, Aircraft Bill of Sale, or its equivalent may be used as evidence of ownership. If the applicant did not purchase the aircraft from the last registered owner, the applicant must submit conveyances completing the chain of ownership from the registered owner to the applicant.

The purchaser under a CONTRACT OF CONDITIONAL SALE is considered the owner for the purpose of registration and the contract of conditional sale must be submitted as evidence of ownership.

A corporation which does not meet citizenship requirements must submit a certified copy of its certificate of incorporation.

REGISTRATION AND RECORDING FEES: The fee for issuing a certificate of aircraft registration is $5; therefore, a $5 fee should accompany this application. An additional $5 recording fee is required when a conditional sales contract is submitted as evidence of ownership. There is no recording fee for a bill of sale submitted with the application.

MAILING INSTRUCTIONS: Please send the WHITE original and GREEN copy of this application to the Federal Aviation Administration Aircraft Registry, Mike Monroney Aeronautical Center, P.O. Box 25504, Oklahoma City, Oklahoma 73125. Retain the pink copy after the original application, fee, and evidence of ownership have been mailed or delivered to the Registry. When carried in the aircraft with an appropriate current airworthiness certificate or a special flight permit, this pink copy is temporary authority to operate the aircraft.

CHANGE OF ADDRESS: An aircraft owner must notify the FAA Aircraft Registry of any change in permanent address. This form may be used to submit a new address.

Fig. 5-3. 8050-1, Aircraft Registration Information.

UNITED STATES OF AMERICA DEPARTMENT OF TRANSPORTATION
FEDERAL AVIATION ADMINISTRATION-MIKE MONRONEY AERONAUTICAL CENTER
AIRCRAFT REGISTRATION APPLICATION

CERT. ISSUE DATE

UNITED STATES
REGISTRATION NUMBER **N**

AIRCRAFT MANUFACTURER & MODEL

AIRCRAFT SERIAL No.

FOR FAA USE ONLY

TYPE OF REGISTRATION (Check one box)

☐ 1. Individual ☐ 2. Partnership ☐ 3. Corporation ☐ 4. Co-owner ☐ 5. Gov't. ☐ 8. Non-Citizen Corporation

NAME OF APPLICANT (Person(s) shown on evidence of ownership. If individual, give last name, first name, and middle initial.)

TELEPHONE NUMBER: (   )

ADDRESS (Permanent mailing address for first applicant listed.)

Number and street: _____

Rural Route: P.O. Box:

CITY	STATE	ZIP CODE

☐ **CHECK HERE IF YOU ARE ONLY REPORTING A CHANGE OF ADDRESS**
ATTENTION! Read the following statement before signing this application.
This portion MUST be completed.

A false or dishonest answer to any question in this application may be grounds for punishment by fine and / or imprisonment (U.S. Code, Title 18, Sec. 1001).

CERTIFICATION

I/WE CERTIFY:

(1) That the above aircraft is owned by the undersigned applicant, who is a citizen (including corporations) of the United States.

(For voting trust, give name of trustee: _____), or:

CHECK ONE AS APPROPRIATE:

a. ☐ A resident alien, with alien registration (Form 1-151 or Form 1-551) No. _____

b. ☐ A non-citizen corporation organized and doing business under the laws of (state) _____
and said aircraft is based and primarily used in the United States. Records or flight hours are available for inspection at _____

(2) That the aircraft is not registered under the laws of any foreign country; and

(3) That legal evidence of ownership is attached or has been filed with the Federal Aviation Administration.

NOTE: If executed for co-ownership all applicants must sign. Use reverse side if necessary.

TYPE OR PRINT NAME BELOW SIGNATURE

EACH PART OF THIS APPLICATION MUST BE SIGNED IN INK.

SIGNATURE	TITLE	DATE
SIGNATURE	TITLE	DATE
SIGNATURE	TITLE	DATE

NOTE Pending receipt of the Certificate of Aircraft Registration, the aircraft may be operated for a period not in excess of 90 days, during which time the PINK copy of this application must be carried in the aircraft.

AC Form 8050-1 (3/90) (0052-00-628-9006) Supersedes Previous Edition

Fig. 5-4. 8050-1, Aircraft Registration Application. (Keep the pink copy as your temporary registration.

THIS FORM SERVES TWO PURPOSES

PART I acknowledges the recording of a security conveyance covering the collateral shown.
PART II is a suggested form of release which may be used to release the collateral from the terms of the conveyance.

PART I – CONVEYANCE RECORDATION NOTICE

NAME (last name first) OF DEBTOR

NAME and ADDRESS OF SECURED PARTY/ASSIGNEE

NAME OF SECURED PARTY'S ASSIGNOR (if assigned)

Do Not Write In This Block
FOR FAA USE ONLY

FAA REGISTRA-TION NUMBER	AIRCRAFT SERIAL NUMBER	AIRCRAFT MFd. (BUILDER) and MODEL

ENGINE MFR. and MODEL	ENGINE SERIAL NUMBER(S)

PROPELLER MFR. and MODEL	PROPELLER SERIAL NUMBER(S)

THE SECURITY CONVEYANCE DATED_____COVERING THE ABOVE COLLATERAL WAS RECORDED BY THE FAA AIRCRAFT REG-ISTRY ON_____ AS CONVEYANCE NUMBER_____.

FAA CONVEYANCE EXAMINER

PART II – RELEASE – (This suggested release form may be executed by the secured party and returned to the FAA Aircraft Registry when terms of the conveyance have been satisfied. See below for additional information.)

THE UNDERSIGNED HEREBY CERTIFIES AND ACKNOWLEDGES THAT HE IS THE TRUE AND LAWFUL HOLDER OF THE NOTE OR OTHER EVIDENCE OF INDEBTEDNESS SECURED BY THE CONVEYANCE REFERRED TO HEREIN ON THE ABOVE-DESCRIBED COLLATERAL AND THAT THE SAME COLLATERAL IS HEREBY RELEASED FROM THE TERMS OF THE CONVEYANCE. ANY TITLE RETAINED IN THE COLLATERAL BY THE CONVEYANCE IS HEREBY SOLD, GRANTED, TRANS-FERRED, AND ASSIGNED TO THE PARTY WHO EXECUTED THE CONVEYANCE, OR TO THE ASSIGNEE OF SAID PARTY IF THE CONVEYANCE SHALL HAVE BEEN ASSIGNED: PROVIDED, THAT NO EXPRESS WARRANTY IS GIVEN NOR IMPLIED BY REASON OF EXECUTION OR DELIVERY OF THIS RELEASE.

This form is only intended to be a suggested form of release, which meets the recording requirements of the Federal Aviation Act of 1958, and the regulations issued thereunder. In addition to these requirements, the form used by the security holder should be drafted in accordance with the pertinent provisions of local statutes and other applicable federal statutes. This form may be reproduced. There is no fee for recording a release. Send to FAA Aircraft Reg-istry, P. O. Box 25504, Oklahoma City, Oklahoma 73125.

ACKNOWLEDGEMENT (If Required By Applicable Local Law):

DATE OF RELEASE: ....

....
(Name of security holder)

SIGNATURE (in ink) ....

TITLE ....

(A person signing for a corporation must be a corporate officer or hold a managerial position and must show his title. A person signing for another should see Parts 47 and 49 of the Federal Aviation Regulations (14 CFR).

Fig. 5-5. 8050-41, Release of Lien.

Federal Communications Commission Form 404, Application for Aircraft Radio Station License (Figs. 5-6 through 5-8), must be completed if you have any radio trans-mitting equipment on board. The tear-off section will remain in the airplane as tempo-rary authorization until the new license is sent to you.

Most forms sent to the FAA or FCC will result in the issuance of a document to you. Be patient; it all takes time, which is the reason for keeping portions of the application forms (aircraft registration pink copy and FAA 404 Temporary Operating Authority).

Approved by OMB
3060-0040
Expires 7/31/94
See below for public
burden estimate

APPLICATION FOR AIRCRAFT RADIO STATION LICENSE

Public reporting burden for this collection of information is estimated to average twenty minutes per response, including the time for reviewing instructions, searching existing data sources, gathering and maintaining the data needed, and completing and reviewing the collection of information. Send comments regarding this burden estimate or any other aspect of this collection of information, including suggestions for reducing the burden to Federal Communications Commission, Information Resources Branch, Room 416, Washington, DC 20554, and to the Office of Management and Budget, Office of Information and Regulatory Affairs, Paperwork Reduction Project (3060-0040), Washington, DC 20503.

GENERAL INFORMATION

RULES AND REGULATIONS

Before preparing this application, refer to FCC Rules, Part 87, "Aviation Services". Contact the U.S. Government Printing Office, Washington, DC 20402, telephone (202) 783-3238 for the correct price.

CORRECT FORM

Use FCC 404 to apply for:

● A new station license when the station aboard an aircraft is first licensed or the ownership of the aircraft is changed and the previous owner is not to continue as the licensee of the station.
● A modified station license when the licensee remains the same, but the operation is to be different from that provided in the license. If the licensee's name or mailing address changes, notify the Commission by letter, see FCC Rules, Part 87.

To renew your license use FCC 405-B which is normally sent to each licensee at the address of record approximately 60 days prior to license expiration. If you have not received FCC 405-B, you may use FCC 404.

Do not use FCC 404 when applying for transmitters or radio frequencies in radio services other than the aviation services (e.g., Amateur, Industrial) even though these facilities may be placed aboard the aircraft. FCC 404 cannot be used to file for a Ground Radio Station License or Restricted Radiotelephone Operator's Permit.

NUMBER OF APPLICATIONS

Submit a separate application form for each aircraft and for each portable radio (see COMPLETING THE APPLICATION, Item 14), unless the application is for a fleet license. See FCC Rules, Part 87 for those eligible for a fleet license.

FEES AND MAILING INSTRUCTIONS

Each application must be accompanied by a single check or money order payable to the FCC for the Total Fee Due. Mail your application and fee to: FEDERAL COMMUNICATIONS COMMISSION, AVIATION AIRCRAFT SERVICE, P. O. BOX 358280, PITTSBURGH, PA 15251-5280.

FEE EXEMPTIONS: No fee is required for governmental entities. Fee exempt applications should be mailed to: FEDERAL COMMUNICATIONS COMMISSION, 1270 FAIRFIELD ROAD, GETTYSBURG, PA 17325-7245.

COMPLETING THE APPLICATION

ITEM 1. Enter the legal name of the person or entity applying for the license. If you are an individual doing business in your own name, enter your full individual name, (last name, first name, and middle initial).

EXAMPLE: Smith, John A.

If you are an individual doing business under a firm or trade name (sole proprietorship), enter both your name and the firm or trade name.

EXAMPLE: Doe, John H. DBA Doe Construction Co.

Do not apply in the name of more than one individual, except on behalf of a legally recognized partnership. If the applicant is a partnership, list the name of the partner whose address appears in items 2 through 6. List the other partners in item 15. If you are a member of a partnership doing business under a firm or company name, insert the full name of each partner having an interest in the business and the firm or company trade name.

EXAMPLE: Doe, John H. & Doe, Richard A. DBA

Doe Construction Company

If you are filing as a corporation, insert the exact name of the corporation as it appears in the Articles of Incorporation. If you are an unincorporated association,

insert the name of the association as it appears in the Articles of Association or By-Laws. If you are a governmental entity, insert the name of the Government entity having jurisdiction of the station.

EXAMPLE: State of California   City of Houston, TX
County of Fairfax, VA

ITEMS 2-6. Enter a permanent mailing address in the United States to which the authorization and any future correspondence related to your station is to be mailed.

ITEM 7. The FAA registration number must be entered on applications submitted for a new station license except those for which FAA registration is not required or those for a fleet or portable license. If exempt from FAA registration, provide an explanation in item 15. When a fleet or portable license is involved, a control number will be assigned by the Commission. When applying for modification or renewal of an existing aircraft radio station license, the FAA registration number or the control number appearing on the license must be entered in item 7.

GOVERNMENTAL ENTITIES ARE EXEMPT FROM FEE REQUIREMENTS AND SHOULD SKIP ITEMS 8 THROUGH 10 OF THE APPLICATION.

FCC 404 INSTRUCTIONS
SEPTEMBER 1991

(CONTINUED ON REVERSE)

Fig. 5-6. FCC 404, Application for Aircraft Radio Station License general information.

COMPLETING THE APPLICATION (CONTINUED)

ITEM 8. Refer to the Private Radio Services Fee Filing Guide for the appropriate Fee Type Code to enter for this application.

ITEM 9. Enter the number of aircraft to be licensed as the Fee Multiple. Normally, the Fee Multiple will be "1", unless the application is for a fleet license, in which case you must show the number of aircraft in the fleet for a new station license, or the number to be added if application is for a modification.

ITEM 10. Refer to the Private Radio Services Fee Filing Guide to determine the fee amount associated with the Fee Type Code in item 8. Multiply the fee amount by the Fee Multiple in item 9, enter the result in item 10, Fee Due. Your check or money order should be for this amount. We will not accept multiple checks.

ITEM 11. Check only one block for the appropriate type of applicant.

ITEM 12. Check the appropriate block for the purpose of filing this application, if for a modification, briefly explain proposed modifications.

ITEM 13. Indicate if application is for a fleet license, if "YES", show the total number of aircraft for a new fleet license, or show the number of aircraft being added or deleted for a modification.

ITEM 14. Check the desired frequencies based on the following information:

PRIVATE AIRCRAFT: These frequencies include those normally available for air traffic control, aeronautical advisory, aeronautical multicom, ground traffic control, and navigation. Refer to Part 87 of the Rules for the specific frequencies available. Private aircraft frequencies are avail-able to any aircraft except those weighing more than 12,500 pounds which are used in carrying passengers or cargo for hire. Do not apply for private aircraft frequencies if the aircraft falls within the latter category.

AIR CARRIER: Refer to Part 87 of the Rules for specific frequencies available.

DO NOT CHECK BOTH PRIVATE AIRCRAFT AND AIR CARRIER IN ITEM 14A.

FLIGHT TEST HF OR VHF OR BOTH: Submit a statement showing that the applicant is a manufacturer of aircraft or major aircraft components. Any request for VHF flight test frequencies must be accompanied by AFTRCC Coordination.

PORTABLE: Submit a statement that it is necessary for the applicant to move the transmitting equipment aboard various U.S. registered aircraft. NOTE: No license is required for a portable radio used only as a back-up on an aircraft which has a station license.

OTHER: Specify any other frequencies you require that are not regularly available for use in accordance with the provisions of Part 87 of the Rules. Each request for "Other" frequencies must be accompanied by a statement showing the need for assignment, including reference to any governmental contracts which may be involved and a description of the proposed use. The emission, power, points of communication, and area of operation should also be included in the statement. In certain cases, AFTRCC Coordination is required.

Application must bear an original signature. Failure to sign the application may result in dismissal of the application and forfeiture of any fees paid.

FCC 404 INSTRUCTIONS
SEPTEMBER 1991

 DETACH HERE -

UNITED STATES OF AMERICA
FEDERAL COMMUNICATIONS COMMISSION

Approved by OMB
3060-0040
Expires 7/31/94
See instructions for
public burden estimate.

TEMPORARY AIRCRAFT RADIO STATION OPERATING AUTHORITY

Use this form if you want temporary operating authority while your regular application, FCC 404, is being processed by the FCC. This authority authorizes the use of transmitters operating on the appropriate frequencies listed in Part 87 of the Commission's Rules.

DO NOT use this form if you already have a valid aircraft station license.
DO NOT use this form when renewing your aircraft license.
DO NOT use this form if you are applying for a fleet license.
DO NOT use this form if you do not have an FAA Registration Number.

ALL APPLICANTS MUST CERTIFY:

1. I am not a representative of a foreign government.
2. I have applied for an Aircraft Radio Station License by mailing a completed FCC Form 404 to the FCC.
3. I have not been denied a license or had my license revoked by the FCC.
4. I am not the subject of any adverse legal action concerning the operation of a radio station license.
5. I ensure that the Aircraft Radio Station will be operated only by individuals properly licensed or otherwise permitted by the Commission's Rules.

WILLFUL FALSE STATEMENTS MADE ON THIS FORM ARE PUNISHABLE BY FINE AND/OR IMPRISONMENT (U.S. CODE, TITLE 18, SECTION 1001), AND/OR REVOCATION OF ANY STATION LICENSE OR CONSTRUCTION PERMIT (U.S. CODE, TITLE 47, SECTION 312(A)(1)), AND/OR FORFEITURE (U.S. CODE, TITLE 47, SECTION 503).

Name of Applicant (Print or Type)	Signature of Applicant
FAA Registration Number (Use as Temporary Call Sign)	Date FCC 404 Mailed

Your authority to operate your Aircraft Radio Station is subject to all applicable laws, treaties and regulations and is subject to the right of control of the Government of the United States. This authority is valid for 90 days from the date FCC 404 is mailed to the FCC.

YOU MUST POST THIS TEMPORARY OPERATING AUTHORITY ON BOARD YOUR AIRCRAFT

FCC 404-A
September 1991

Fig. 5-7. FCC 404, Application for Aircraft Radio Station License Temporary Operating Authority. (Tear off bottom and retain until license comes in the mail.)

Approved by OMB
3060-0040
Expires 7/31/94
See instructions for
public burden estimate.

UNITED STATES OF AMERICA
FEDERAL COMMUNICATIONS COMMISSION

FOR
FCC
USE
ONLY

APPLICATION FOR AIRCRAFT RADIO STATION LICENSE

1. APPLICANT NAME

2. MAILING ADDRESS (Line 1)

3. MAILING ADDRESS (Line 2)

4. CITY

5. STATE	6. ZIP CODE	7. FAA REGISTRATION OR FCC CONTROL NUMBER (If FAA registration is not required for your aircraft, explain in item 15) N_____	
8. FEE TYPE CODE	9. FEE MULTIPLE	10. FEE DUE $	FOR FCC USE ONLY

11. TYPE OF APPLICANT

☐ I—Individual

☐ D—Individual with Business Name

☐ P—Partnership

☐ C—Corporation

☐ A—Association

☐ G—Governmental Entity

12. PURPOSE OF APPLICATION

☐ New Station   ☐ Renewal

☐ Modification (Specify) _____

_____

13. IS APPLICATION FOR A FLEET LICENSE?   ☐ YES   ☐ NO

A. If modifying a fleet license, give the number of aircraft to be added.

B. If applying for a new or modified fleet license, give the total number of aircraft.

14. FREQUENCIES REQUESTED (Check appropriate box(es) in 14A and/or 14B, see Instructions)

A. CHECK ONLY ONE

☐ A—Private Aircraft

☐ C—Air Carrier

B. ADDITIONAL INFORMATION IS REQUIRED IF YOU CHECK HERE

☐ T—Flight Test HF   ☐ P—Portable (Showing required)

☐ V—Flight Test VHF   ☐ O—Other (Specify) _____

15. ANSWER SPACE FOR ADDITIONAL INFORMATION

CERTIFICATION

1. Applicant waives all claims for the use of any specific frequency regardless of prior use by license or otherwise.
2. Applicant will have unlimited access to the radio equipment and will control access to exclude unauthorized persons.
3. Neither applicant nor any member thereof is a foreign government or representative thereof.
4. Applicant certifies that all statements made in this application and attachments are true, complete, correct and made in good faith.
5. Applicant certifies that the signature that appears on this application is that of a person with the proper authority to act on behalf of the party represented.

WILLFUL FALSE STATEMENTS MADE ON THIS FORM ARE PUNISHABLE BY FINE AND/OR IMPRISONMENT (U.S. CODE, TITLE 18, SECTION 1001), AND/OR REVOCATION OF ANY STATION LICENSE OR CONSTRUCTION PERMIT (U.S. CODE, TITLE 47, SECTION 312(A)(1)), AND/OR FORFEITURE (U.S. CODE, TITLE 47, SECTION 503).

⇨ SIGNATURE   DATE

FAILURE TO SIGN THIS APPLICATION MAY RESULT IN DISMISSAL OF THE APPLICATION AND FORFEITURE OF ANY FEES PAID.

FCC 404
September 1991

Fig. 5-8. FCC 404, Application for Aircraft Radio Station License.

Assistance

Many forms must be completed and, although they are not complicated, you may seek assistance to fill them out. Check with an FBO or call upon another party, such as AOPA or the lending institution that might be "carrying the note."

AOPA, for a small fee, will provide closing services via telephone, plus prepare and file the necessary forms to complete the transaction. This is particularly convenient if the parties involved in the transaction are spread all over the country, as would be the case if you are purchasing an airplane sight unseen.

INSURANCE

Insure your airplane from the moment you sign on the dotted line (a lending institution will most likely require insurance to protect its financial stake). No one can afford to take risks. Two types of insurance are *liability* and *hull*.

Liability insurance protects you, or your heirs, in instances of claims (bodily injury or property damage) against you resulting from your operation of an airplane. Expect a lawsuit if someone is injured or killed as a result of your flying.

Hull insurance protects your investment from loss caused by the elements of nature, by fire, by theft, by vandalism, or while being operated. Limited coverages are available that will cover losses to the airplane while on the ground, but not while in the air. You can save money here; however, discussion of coverages available is best left to you and your insurance agent. Your lending institution will require hull insurance for their protection.

Aviation publications typically have telephone numbers for several aviation underwriters and many of the underwriters have a toll-free telephone number. Check with more than one company because services, coverage, and rates do differ. Stay clear of policies that have exclusions or other specific rules involving maximum preset values for replacement parts.

Excellent sources of insurance information are the insurance companies sponsored by the Cessna Owners Organizations and the Cessna Pilots Association. Addresses and phone numbers for both appear in chapter 12. Review your health and life insurance coverage to make sure you are covered while flying a small airplane.

PROTECTING YOUR AIRPLANE

No airplane can be made theftproof; however, the airplane can be less attractive to the thief. Less attractive means more difficult to enter, damage, or steal.

Prevention

No thief wants to spend a long time entering an airplane: get in and go without delay, just that fast. Any delaying device might discourage the thief and prevent a crime. Locks are the most well-accepted time-delay devices known and are used to prevent entry to everything. Storing the airplane in a locked hangar is the best method. Or use cut-resistance chain and locks for tiedowns. Install a throttle or control lock. Store the airplane inside a locked perimeter fence.

A remote alarm system can be installed that could catch the thief in the act; however, quick response by responsible personnel is required. An internal alarm is available that makes the interior of the cabin a most unpleasant place to stay by sounding an inordinately loud audio signal from an inaccessible speaker. For information contact:

Thompson Burglar Alarm Systems
36516 N. Echo Road
Deer Park, WA 99006
(509) 292-2162

REPORTING A STOLEN AIRPLANE

Suppose you drive out to the airport and your airplane had been stolen, or the plane has been broken into and some of your avionics were missing. What do you do?

1. Notify the law enforcement authority responsible for the airport: airport police, city police, county police, sheriff, or state police. Someone might come to the airport and make a report of the theft, perhaps even processing the crime scene by obtaining any fingerprints and taking photographs.
2. Notify the FAA, or demand that the police do so, to facilitate the issuance of a nationwide stolen aircraft alert.
3. Notify the ACPI (Aviation Crime Prevention Institute) at (301) 694-5444 or FAX (301) 695-6955.
4. Notify your insurance company of the loss, and be ready to supply them with copies of all police reports, purchase receipts, and the like.

Don't expect the responsible law enforcement agency to do very much about the theft. Reports will be filed and entries will be made into the National Crime Information Center (NCIC) computer including registration numbers, serial numbers, and the like, to improve the chance of recovery in the event that another police department on the other side of the country locates the airplane.

One of the strongest recommendations is to have excellent records of all equipment installed or stored on board the airplane. Photographs also are advised. This type of evidence will aid you in settling claims with your insurance company.

6

Inspections

TO CARE FOR YOUR AIRPLANE and assure its continued airworthiness, it is required that certain items be inspected for proper operation and integrity. Although sounding very official and legalistic, this statement shows the reasoning for inspections.

General inspection requirements in this chapter are excerpted and edited from the *Cessna 172 Service Manual*. This chapter should help you come to understand why a proper annual takes time and costs money. Don't cut yourself short with a shoddy inspection. You may save time and money, but it could cost you your life.

To avoid repetition throughout the inspection, general points to be checked are given below. In the inspection, only the items to be checked are listed; details as to how to check, or what to check for, are excluded. The inspection covers several different models. Some items might apply only to specific models, and some items are optional equipment that might not be found on a particular airplane. Check the FAA airworthiness directives and Cessna service letters for compliance at the time specified by them. Federal Aviation Regulations require that all civil aircraft have a periodic (annual) inspection as prescribed by the Administrator, and performed by a person designated by the Administrator. The Cessna Aircraft Company recommends a 100-hour periodic inspection for the airplane.

CHECK AS APPLICABLE

Movable parts: lubrication, servicing, security of attachment, binding, excessive wear, safetying, proper operation, proper adjustment, correct travel, cracked fittings, security of hinges, defective bearings, cleanliness, corrosion, deformation, sealing, and tensions.

Fluid lines and hoses: leaks, cracks, dents, kinks, chafing, proper radius, security, corrosion, deterioration, obstructions, and foreign matter.

Metal parts: security of attachment, cracks, metal distortion, broken spotwelds, corrosion, condition of paint, and any other apparent damage.

Wiring: security, chafing, burning, defective insulation, loose or broken terminals, heat deterioration, and corroded terminals.

Bolts in critical areas: correct torque in accordance with proper torque values, when installed or when visual inspection indicates the need for a torque check.

Filters, screens, and fluids: cleanliness, contamination and/or replacement at specified intervals.

AIRPLANE FILE

Miscellaneous data, information, and licenses are a part of the airplane file. Check that the following documents are up-to-date and in accordance with current FARs. Most of the items listed are required by FARs. Because the regulations of other nations might require other documents and data, owners of exported aircraft should check with their aviation regulatory officials to determine specific requirements.

To be displayed in the airplane at all times: aircraft airworthiness certificate, aircraft registration certificate, and aircraft radio license.

To be carried in the airplane at all times: weight and balance and associated papers, plus the aircraft equipment list.

To be made available upon request: airframe logbook and engine logbook.

ENGINE RUN-UP

Before beginning the step-by-step inspection, start, run up, and shut down the engine in accordance with instructions in the owner's manual. During the run-up, observe the following, making note of any discrepancies or abnormalities:

- Engine temperatures or pressures
- Static rpm
- Magneto drop
- Engine response to changes in power
- Any unusual engine noises
- Fuel selector valve; operate the engine on each position long enough to make sure the valve functions properly
- Idling speed and mixture; proper idle cutoff.
- Alternator
- Suction gauge

After the inspection has been completed, an engine run-up should be performed again to verify that problems have been corrected.

PERIODIC INSPECTIONS

The items that follow are specified for service based upon time in use (hours of operation).

Engine lubrication

Continental engine. If the engine is equipped with an external oil filter, change the engine oil and filter element at 50-hour intervals. If the engine is not equipped with an external oil filter, change the engine oil and clean the oil screen every 25 hours.

Lycoming engine. If the engine is not equipped with an external oil filter, change the engine oil and clean the oil screens at 50-hour intervals. If the engine is equipped with an external oil filter, the engine oil change intervals can be extended to 100-hour intervals, providing the external filter element is changed at 50-hour intervals.

The 50-hour inspection includes a visual check of the engine, propeller, and aircraft exterior for any apparent damage or defects; an engine oil change as required above; and accomplishment of lubrication and servicing requirements. Remove the propeller spinner and engine cowling, then replace after the inspection has been completed.

The 100-hour or annual inspection includes everything in the 50-hour inspection, and oil change as required above. Also loosen or remove the fuselage, wing, empennage, and upholstery inspection doors, plates, and fairings only as necessary to perform a thorough, searching inspection of the aircraft. Replace after the inspection has been completed.

Note: In the following charts, numbers indicate the time in hours between inspections or servicing.

Propeller

1. Spinner and spinner bulkhead—50
2. Blades—50
3. Hub—50
4. Bolts and/or nuts—50

Engine compartment

Check for evidence of oil and fuel leaks, then clean the entire engine and compartment, if needed, prior to inspection.

1. Engine oil, screen, filler cap, dipstick, drain plug, and filter—50
2. Oil cooler—50
3. Induction air filter—50
4. Induction air box, air valves, doors, and controls—50
5. Cold and hot air hoses—50
6. Engine baffles—50
7. Cylinders, rocker box covers, and push rod housings—50
8. Crankcase, oil sump, accessory section, and front crankshaft seal—50
9. All lines and hoses—50
10. Intake and exhaust systems—50
11. Ignition harness—50
12. Spark plugs and compression—100
13. Crankcase and vacuum system breather lines—50
14. Electrical wiring—50
15. Vacuum pump, oil separator, and relief valve—50
16. Vacuum relief valve filter—100
17. Engine controls and linkage—50
18. Engine shock mounts, engine mount structure, and ground straps—50

19. Cabin heater valves, doors, and controls—50
20. Starter, solenoid, and electrical connections—50
21. Starter brushes, brush leads, and commutator—200
22. Alternator and electrical connections—50
23. Alternator brushes, brush leads, and commutator or slip ring—500
24. Voltage regulator mounting and electrical leads—50
25. Magnetos (externally) and electrical connections—50
26. Slick magneto timing—100
27. Carburetor—50
28. Firewall—100
29. Engine cowling—50
30. Carburetor drain plug for security—50

Fuel system

1. Fuel strainer, drain valve, and control—50
2. Fuel strainer screen and bowl—100
3. Fuel tanks, fuel lines, drains, filler caps, and placards—100
4. Drain fuel and check tank interior, attachment, and outlet screens—1000
5. Fuel vents and vent valves—100
6. Fuel selector valve and placards—100
7. Engine primer—100

Landing gear

1. Brake fluid, lines and hoses, linings, disc, brake assemblies, and master cylinders—100
2. Main gear wheels, wheel bearings, step and spring strut, tires, and fairings—100
3. Main and nose gear wheel bearing lubrication—500
4. Torque link lubrication—50
5. Nose gear strut servicing—100
6. Nose gear shimmy damper service—100
7. Nose gear wheels, wheel bearings, strut, steering system, shimmy damper, tire, fairing, and torque links—100
8. Parking brake system—100

Airframe

1. Aircraft exterior—50
2. Aircraft structure—100
3. Windows, windshield, and doors—50
4. Seats, stops, seat rails, upholstery, structure, and seat mounting—50
5. Safety belts and attaching brackets—50
6. Control U-bearings, sprockets, pulleys, cables, chains, and turnbuckles—100
7. Control lock, control wheel, and control U mechanism—100
8. Instruments and markings—100
9. Gyros, central air filter—100
10. Magnetic compass compensation—1000

11. Instrument wiring and plumbing—100
12. Instrument panel, shock mounts, ground straps, cover, and decals and labeling—100
13. Defrosting, heating, and ventilating systems, and controls—100
14. Cabin upholstery, trim, sun visors, and ashtrays—100
15. Area beneath floor, lines, hoses, wires, and control cables—100
16. Lights, switches, circuit breakers, fuses, and spare fuses—50
17. Exterior lights—50
18. Pitot and static system—100
19. Stall warning sensing unit, pitot warning heater—100
20. Radio and radio controls—100
21. Radio antennas—100
22. Battery, battery box, and battery cables—100
23. Battery electrolyte level—50

Control systems

In addition to the items listed below, always check for correct direction of movement, correct travel, and correct cable tension.

1. Cables, terminals, pulleys, pulley brackets, cable guards, turnbuckles, and fairleads—100
2. Chains, terminals, sprockets, and chain guards—100
3. Trim control wheel, indicators, and actuator—100
4. Travel stops—100
5. All decals and labeling—100
6. Flap control switch (or lever), flap rollers and tracks, flap transmitter and linkage, and flap position indicator, flap electric motor and transmission—100
7. Elevator trim—100
8. Rudder pedal assemblies and linkage—100
9. Skin and structure of control surface and trim tabs—100
10. Balance weight attachment—100

Note: A high-time inspection is merely a 100-hour inspection with the addition of an engine overhaul. At the time of overhaul, engine accessories should be overhauled.

PROPELLERS

Most pilots give very little thought to their fixed pitch metal propeller. They should because even though a high margin of safety is designed into the typical modern propeller, failures do occur. It is fair to point out that the propeller is at the end of the energy chain, is responsible for efficiently converting the brake horsepower of the engine into thrust, and is designed to be tough.

Studying reports of propeller blade failure does not indicate that failures are more prevalent among specific aircraft and engine and powerplant combinations. Failures occur with all types of propellers: fixed-pitch, ground-adjustable, variable-pitch, and constant-speed.

Although most 172s are equipped with fixed-pitch propellers, a few have constant-speed props. Note that the constant-speed propeller can consist of as many as 200 parts.

Investigations into a number of blade failures show that the majority of failures are due to fatigue cracks that started from mechanically formed dents (damage), cuts, scars, scratches, nicks, or leading edge pits. Only rare cases indicated failure caused by material defects (surface discontinuities or internal) existing from the time of manufacture.

Fatigue failure and flutter

Fatigue failure generally happens at the location of previous propeller repairs. The failure might have started as a result of the initial damage or from improperly performed repairs. Note that blade-straightening or blade-pitch modifications can overstress the metal, causing it to fail.

Flutter, a vibration causing the ends of the blade to twist back and forth around an axis perpendicular to the crankshaft, can cause propeller blade failure (by metal fatigue).

Certain engine speeds can cause the vibration to become critical. If the propeller (engine) is allowed to operate in this range, propeller blade failure might occur. Such warning or limitations are to be found in operating manuals or marked on the tachometer; therefore, tachometer accuracy is very important and periodic tachometer accuracy checks should be made.

Stress

Normal stresses that occur in propeller blades can be visualized as parallel lines of force within the blade, running approximately parallel to the surface. During normal operation, the four separate stresses imposed upon a propeller are: thrust, torque, centrifugal force, and aerodynamic force. Other stresses might be imposed by fluttering or uneven tracking of the blades.

When a defect occurs on the blade, the defect tends to squeeze these lines of force together, concentrating the stress into a small area. The increased stress might be sufficient to form a crack at that location. The crack then causes a greater stress concentration resulting in growth of the crack. This scenario repeats until blade failure.

Under most circumstances, fatigue failures of propeller blades occur within a few inches of the blade tip; however, failures can happen at any location where dents, cuts, scratches, or nicks are ignored; therefore, no damage should be overlooked or allowed to go without repair.

Inspection

Always act as if the propeller is "hot" and the engine capable of starting. Sometimes only a small propeller movement can start an engine.

When inspecting a propeller during the preflight, check the entire blade—not just the leading edge—for erosion, scratches, nicks, dents, and cracks. Regardless of how small or minor any surface irregularities might appear, consider them as points of stress and subject to fatigue failure.

Propeller manufacturer's manuals, service letters, and bulletins specify methods and limits for blade maintenance, inspection, service, and repair. When repairs to a damaged blade are possible, the manufacturer's instructions must be followed using accepted industry practices and techniques. All propeller repairs must be performed by qualified repair agencies and competent personnel.

Pilots and owners must be aware of airworthiness directives that apply to a propeller and pay close attention to the repetitive requirements of any applicable ADs.

Propeller care

The following tips will help you care for the propeller:

- Keep blades clean because dirt, oil, grease, and the like, can mask cracks or other defects.
- Avoid engine run-up areas containing loose sand, stones, gravel, or broken asphalt because these particles can scar your propeller.
- The propeller is not a handle for moving the airplane. Not only can the propeller be damaged, but the engine could accidently be started causing death, injury, and property damage.
- The engine tachometer must be checked for accuracy to preclude operation in any restricted rpm range.
- The best way to prolong the life of a propeller is regular care and maintenance.

7

Caring for a 172

WHEN YOU OWN AN AIRPLANE, you are responsible for its care. An owner needs to be concerned about four areas of care: proper handling, effective storage, cleaning, and preventive maintenance. All four help the owner become intimately acquainted with the airplane. Additionally, by properly caring for the airplane, the investment will be protected, safety improved, and a considerable amount of money saved. Handling, storage, and cleaning are purely matters of personal preference. Preventive maintenance is a matter of federal regulation; thus, preventive maintenance is the topic of chapter 8.

GROUND HANDLING

Proper handling of an airplane (moving by hand) while on the ground is extremely important. If care is not taken, major structural damage can be done to the airplane that could cost thousands of dollars to repair. A tow bar properly attached to the nose gear should be used for pulling, pushing, steering, and maneuvering the aircraft while on the ground (Figs. 7-1 and 7-2). To prevent damage to the nose gear, never turn the nosewheel more than 30° either side of center.

If no tow bar is available, apply downward pressure at the horizontal stabilizer front spar adjacent to the fuselage to raise the nosewheel off the ground. When the nosewheel is clear of the ground, the aircraft can be maneuvered by pivoting the main wheels. This is not the recommended procedure, but can be used in a pinch. The best method is to use a tow bar.

If more pushing or pulling power is needed when moving the aircraft with the tow bar, use only the wing struts and landing gear legs as push points. Don't push on the wing edges, control surfaces, cabin doors, and the like.

Never use the propeller as a push point. The engine could fire from only a slight movement, causing severe injury or death. If you are macho about that, then think about propeller blade failure caused by bending. Blade failure means complete separation caused by metal fatigue developed from stress caused by bending. I see airplanes pushed and pulled by their propellers all the time, as will you, but don't do it.

Parking procedures will depend upon the local conditions of the airport and the weather. Most pilots will park an airplane for a short period of time by applying the

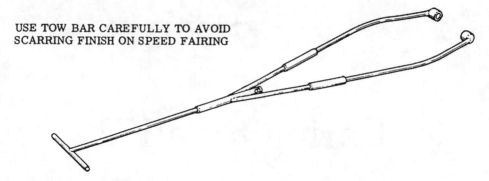

USE TOW BAR CAREFULLY TO AVOID
SCARRING FINISH ON SPEED FAIRING

Fig. 7-1. Old-style tow bar. Cessna Aircraft Company

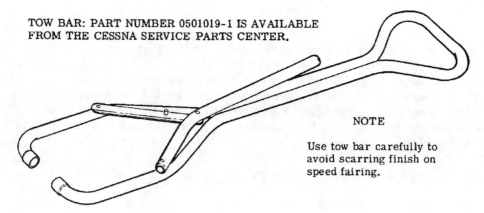

TOW BAR: PART NUMBER 0501019-1 IS AVAILABLE
FROM THE CESSNA SERVICE PARTS CENTER.

NOTE

Use tow bar carefully to
avoid scarring finish on
speed fairing.

Fig. 7-2. New-style tow bar. Cessna Aircraft Company

parking brake and locking the controls. Further caution would call for wheel chocks. This procedure is adequate—when no wind is blowing—for short periods of time such as fueling, pit stops, and other short layovers when the airplane will not be out of sight.

If the airplane is going to be left unattended for more than a few minutes, tie it down (Fig. 7-3). Inside the airplane, lock the control surfaces with the internal control lock and set the brakes. If no standard Cessna control lock is available, tie the controls down with the seat belt. Do not set the parking brakes during cold weather or when the brakes are overheated.

Outside, face the airplane into the wind, then:

1. Attach ropes, cables, or chains to the wing tiedown fittings located at the upper end of each wing strut. Secure the opposite ends to ground anchors.
2. Secure a tiedown rope, not a chain or cable, to the exposed portion of the engine mount and secure the opposite end to a ground anchor.
3. Fasten the middle of a rope to the tail tiedown ring, pull each end of the rope away at a 45° angle, and secure to ground anchors at each side of the tail.
4. Install surface control locks between the wingtip and aileron, and over the fin and rudder.

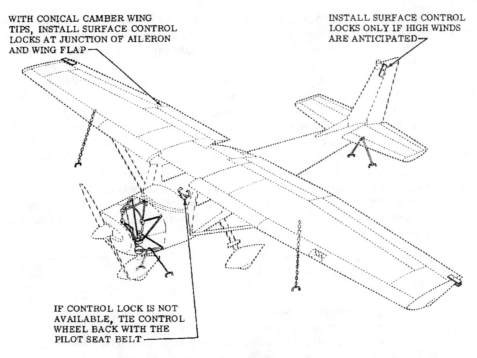

WITH CONICAL CAMBER WING
TIPS, INSTALL SURFACE CONTROL
LOCKS AT JUNCTION OF AILERON
AND WING FLAP —

INSTALL SURFACE CONTROL
LOCKS ONLY IF HIGH WINDS
ARE ANTICIPATED —

IF CONTROL LOCK IS NOT
AVAILABLE, TIE CONTROL
WHEEL BACK WITH THE
PILOT SEAT BELT —

Fig. 7-3. Recommended method of tiedown. Cessna Aircraft Company

Tie the airplane down whether the wind is blowing or not. The airplane might be damaged by unexpected winds, or might damage other airplanes, if you neglect to secure the airplane and leave it unattended. Unexpected winds often visit in the form of dust devils in the summertime. A dust devil is a twisting wind that is quite capable of picking up a 172 and throwing it around.

AIRPLANE STORAGE

Storing an aircraft is much more than hangaring or parking. Unfortunately, a large number of owners pay little heed to the proper preservation of an airplane during periods of nonuse.

Engine storage

(Edited excerpts from a Teledyne Continental Motors (TCM) service bulletin (M84-10 Rev.1) provide excellent information regarding engine storage. According to TCM, the procedures are a general recommendation. Local conditions are different and TCM has no control over the application; therefore, more stringent procedures might be required. Rust and corrosion prevention are the owner's responsibility, TCM says.)

Engines in aircraft that are flown occasionally tend to exhibit cylinder wall corrosion more than engines in aircraft that are flown frequently. Of particular concern are new engines or engines with new or freshly honed cylinders after a top or major overhaul. In areas of high humidity, corrosion has been found in such cylinders after an inactive period of only a few days. When cylinders have been operated for approximately 50 hours, the varnish deposited on the cylinder walls offers some protection against corrosion; hence, a two-step program for flyable storage category is recommended.

Obviously, proper steps must be taken on engines used infrequently to lessen the possibility of corrosion. This is especially true if the aircraft is based near the sea coast or in areas of high humidity and flown less than once a week. In all geographical areas, the best method of preventing corrosion of the cylinders and other internal parts of the engine is to fly the aircraft at least once a week, long enough to reach normal operating temperatures, which will vaporize moisture and other byproducts of combustion.

In consideration of the circumstances mentioned, TCM has developed three reasonable minimum preservation procedures that, if implemented, will minimize the detriments of rust and corrosion. It is the owner's responsibility to choose a program that is viable to the particular aircraft's mission. Aircraft engine storage recommendations are broken down into categories:

    A. Flyable storage (Program I or II)
    B. Temporary storage (up to 90 days)
    C. Indefinite storage

A. Flyable storage (Program I or II)

Program I is for engines or cylinders with fewer than 50 operating hours:

    a. Propeller pull-through every five days as per paragraph A2; and
    b. Fly every 15 days as per paragraph A3.

Program II is for engines or cylinders with more than 50 operating hours to TBO if not flown weekly:

    a. Propeller pull-through every seven days as per paragraph A2; and
    b. Fly every 30 days as per paragraph A3.

1. Service aircraft per normal airframe manufacturer's instructions.
2. The propeller should be rotated by hand without running the engine. For four- and six-cylinder straight-drive engines, rotate the engine six revolutions, then stop the propeller 45 to 90 degrees from the original position. For six-cylinder geared engines, rotate the propeller four revolutions and stop the propeller 30 to 60 degrees from the original position.

   Caution: For maximum safety, accomplish engine rotation as follows:
       a. Verify that the magneto switches are OFF.
       b. Throttle position to CLOSED.
       c. Mixture control to IDLE CUT-OFF.
       d. Set brakes and block the aircraft's wheels.
       e. Leave the aircraft tiedowns installed and verify that the cabin door latch is open.
       f. Do not stand within the arc of the propeller blades while turning the propeller.
3. The aircraft should be flown for thirty (30) minutes, reaching, but not exceeding, normal oil and cylinder temperatures. If the aircraft cannot be flown, it should be represerved in accordance with "B"(Temporary Storage) or "C"(Indefinite Storage). Ground running is not an acceptable substitute for flying.

Note: If the airplane cannot be flown on schedule due to weather, maintenance, and the like, pull the propeller through daily and accomplish as soon as possible. It is necessary that for future reference, if required, the propeller pull-through and flight time be recorded and verified in the engine maintenance record/log with the date, time, and signature.

B. Temporary storage (up to 90 days)

1. Preparation for Storage
   a. Remove the top spark plug and spray preservative oil (Lubrication Oil—Contact and Volatile Corrosion—Inhibited, MIL-L-46002, Grade 1) at room temperature through the upper spark plug hole of each cylinder with the piston approximately in bottom dead center position. Rotate the crankshaft as each pair of opposite cylinders is sprayed. Stop the crankshaft with no piston at top dead center. A pressure pot or pump-up type garden pressure sprayer may be used. The spray head should have ports around the circumference to allow complete coverage of the cylinder walls.
   Note: Listed below are some approved preservative oils—that are available from any FBO—recommended for use in Teledyne Continental engines for temporary and indefinite storage:
   MIL-L-46002, Grade 1 Oils
   NOX RUST VCI-105
   PETROTECT VA
   b. Respray each cylinder without rotating the crank. To thoroughly cover all surfaces of the cylinder interior, move the nozzle or spray gun from the top to the bottom of the cylinder.
   c. Reinstall spark plugs.
   d. Apply preservative to the engine interior by spraying the above specified oil (approximately two ounces) through the oil filler tube.
   e. Seal all engine openings exposed to the atmosphere using suitable plugs, or moisture-resistant taper and attach red streamers at each point.
   f. Engines, with propellers installed, that are preserved for storage in accordance with this section should have a tag affixed to the propeller in a conspicuous place with the following notation on the tag:

   DO NOT TURN PROPELLER—ENGINE PRESERVED; PRESERVATION DATE _____.

   Note: If the engine is not returned to flyable status at the expiration of the temporary (90-day) storage, it must be preserved in accordance with the indefinite storage procedures.
2. Preparation for service
   a. Remove seals, tape, paper, and streamers from all openings.
   b. With bottom spark plugs removed from the cylinders, hand turn the propeller several revolutions to clear excess preservative oil, then reinstall the spark plugs.
   c. Conduct normal start-up procedure.
   d. Give the aircraft a thorough cleaning and visual inspection. A test flight is recommended.

C. Indefinite storage

1. Preparation for Storage
   a. Drain the engine oil and refill with MIL-C-6529 Type II. The aircraft should be flown for thirty (30) minutes, reaching, but not exceeding normal oil and cylinder temperatures. Allow the engine to cool to ambient temperature. Accomplish steps 1.a. and 1.b. of Temporary Storage.
   Note: MIL-C-6529 Type II may be formulated by thoroughly mixing one part compound MIL-C-6529 Type I (Esso Rust-Ban 628, Cosmoline No. 1223 or equivalent) with three parts new lubricating oil of the grade recommended for service (all at room temperature). Single grade oil is recommended.
   b. Apply preservative to the engine interior by spraying MIL-L-46002, Grade 1 oil (approximately two ounces) through the oil filler tube.
2. Install dehydrator plugs MS27215-1 or -2, in each of the top spark plug holes, making sure that each plug is blue in color when installed. Protect and support the spark plug leads with AN-4060 protectors.
3. If the engine is equipped with a pressure-type carburetor, preserve this component by the following method: Drain the carburetor by removing the drain and vapor vent plugs from the regulator and fuel control unit. With the mixture control in rich position, inject lubricating oil, grade 1010, into the fuel inlet at a pressure not to exceed 10 psi until oil flows from the vapor vent opening. Allow excess oil to drain, plug the inlet, and tighten and safety the drain and vapor vent plugs. Wire the throttle in the open position, place bags of desiccant in the intake, and seal the opening with moisture-resistant paper and tape, or a cover plate.
4. If the carburetor is removed from the engine, place a bag of desiccant in the throat of the carburetor air adapter. Seal the adapter with moisture-resistant paper and tape or a cover plate.
5. The TCM fuel injection system does not require any special preservation preparation. For preservation of the Bendix RSA-7DA1 fuel injection system, refer to the Bendix Operation and Service Manual.
6. Place a bag of desiccant in the exhaust pipes and seal the openings with moisture resistant tape.
7. Seal the cold air inlet to the heater muff with moisture resistant tape to exclude moisture and foreign objects.
8. Seal the engine breather by inserting a dehydrator MS27215-2 plug in the breather hose and clamping in place.
9. Attach a red streamer to each place on the engine where bags of desiccant are placed. Either attach red streamers outside of the sealed area with tape or to the inside of the sealed area with safety wire to prevent wicking of moisture into the sealed area.
10. Engines, with propellers installed, that are preserved for storage in accordance with this section should have a tag affixed to the propeller in a conspicuous place with the following notation on the tag:

DO NOT TURN PROPELLER—ENGINE PRESERVED; PRESERVATION DATE _____.

Procedures necessary for returning an aircraft to service are as follows:

1. Remove the cylinder dehydrator plugs and all paper, tape, desiccant bags, and streamers used to preserve the engine.
2. Drain the corrosion preventive mixture and reservice with recommended lubricating oil. **Warning:** When returning the aircraft to service, do not use the corrosion preventive oil referenced in paragraph C.1.a. for more than 25 hours.
3. If the carburetor has been preserved with oil, drain it by removing the drain and vapor vent plugs from the regulator and fuel control unit. With the mixture control in rich position, inject service-type gasoline into the fuel inlet at a pressure not to exceed 10 psi until all of the oil is flushed from the carburetor. Reinstall the carburetor plugs and attach the fuel line.
4. With the bottom plugs removed, rotate the propeller to clear excess preservative oil from the cylinders.
5. Reinstall the spark plugs and rotate the propeller by hand through compression strokes of all the cylinders to check for possible liquid lock. Start the engine in the normal manner.
6. Give the aircraft a thorough cleaning, visual inspection and test flight per airframe manufacturer's instructions.

Inspections

Aircraft stored in accordance with the indefinite storage procedures should be inspected per the following instructions:

1. Aircraft prepared for indefinite storage should have the cylinder dehydrator plugs visually inspected every 15 days. The plugs should be changed as soon as their color indicates unsafe conditions of storage. If the dehydrator plugs have changed color on one-half or more of the cylinders, all desiccant material on the engine should be replaced.
2. The cylinder bores of all engines prepared for indefinite storage should be resprayed with corrosion preventive mixture every six months, or more frequently if a bore inspection indicates corrosion has started earlier than six months. Replace all desiccant and dehydrator plugs. Before spraying, the engine should be inspected for corrosion as follows: Inspect the interior of at least one cylinder on each engine through the spark plug hole. If the cylinder shows the start of rust, spray the cylinder with corrosion preventive oil and turn the prop over six times, then respray all cylinders. Remove at least one rocker box cover from each engine and inspect the valve mechanism.

Airframe storage

Cessna aircraft are constructed of corrosion-resistant Alclad aluminum that will last indefinitely under normal conditions, if kept clean; however, the alloys are subject to oxidation. The first indication of oxidation (corrosion) on unpainted surfaces is the formation of white deposits or spots. On painted surfaces, the paint is discolored or blistered. Storage in a dry hangar is essential to good preservation and should be procured if possible.

Short-term storage (fewer than 60 days)

1. Fill the fuel tanks with the correct grade of fuel.
2. Clean and wax the aircraft
3. Clean any grease/oil from the tires and coat them with a tire preservative. Cover the nosewheel to protect it from oil drips.
4. Block up the fuselage to remove the weight from the tires.
   Note: Tires will take a set, causing them to become out-of-round, if an aircraft is left parked for more than a few days.

Long-term storage (indefinite)

1. Proceed with items a through d under Short-Term Storage.
2. Lubricate all airframe items and seal or cover openings.
3. Cover all openings to the airframe. This is to keep vermin, insects, and birds out.
4. Remove the battery and store it in a cool, dry place. Service it periodically.
5. Place covers over the windshield and rear windows.
6. Inspect the airframe for signs of corrosion at least monthly. Clean and wax as necessary.

Returning to service

1. Remove the aircraft from the blocks and check the tires for proper inflation. Check the nose strut for proper inflation.
2. Remove all covers and plugs, and inspect the interior of the airframe for debris and foreign matter.
3. Check the battery and reinstall.
4. Clean and inspect the exterior of the aircraft.

INTERIOR AND EXTERIOR CARE

Proper care is not only important to preserve appearance and value, but to maintain safety as well. What looks good generally sells well and is safe to operate. The interior of the airplane is seen by all, including the pilot, passengers, and eventually the next owner. It is also used by all and is a problem to keep clean. A very thorough cleaning with standard automobile cleaning methods and similar materials, then coating seats and carpets with ScotchGuard, will improve and protect the appearance of the cabin.

All hard surfaces can be cleaned and protected with products such as Protect All, which is available in spray cans and by bulk from:

Protect All, Inc.
1910 East Via Burton St.
Anaheim, CA 92806
(800) 322-4491
(714) 635-4491

Formula 40 is available as a sealer and restoration overcoat for plastic and vinyl surfaces needing more than just cleaning and protection. It is available in several colors. Contact:

Superflite
2149 E. Pratt Blvd.
Elk Grove Village, IL 60007
(800) 323-0611

Interior heat and sunlight protection

Solid-state electronics in general aviation make pilots concerned about heat buildup within the interior of the parked aircraft. Solar heat buildup will affect avionics, plus cause problems with the instrument panel materials, upholstery, and other plastics. The interior temperature of a parked aircraft can reach as much as 185°F. This is the reason that many aircraft tied down outside have a cover over the windshields, either inside or outside (Figs. 7-4 and 7-5). A quick look around at the local airport will find four methods of heat protection:

- The first method that is seen most often on older, already sad-looking airplanes is no protection at all.

- Some owners might recognize the need for protection from the sun's rays and lay a towel or chart over the top of the instrument panel. This is merely an exercise in futility because no heat buildup is prevented. Do this only when parking for a short period of time, such as refueling.

- Inside covers protect the interior of the aircraft by reflecting the sun's rays away with metallic-type reflective surfaces. The heat shields, as they are called, attach to the interior of the aircraft with fasteners. Interior reflective heat shields are available from many sources and are advertised in all the aviation periodicals.

- Exterior airplane covers provide similar protection for the interior of the aircraft, yet give additional exterior protection by covering the windshield, refueling caps, and fresh air vents. Exterior covers are also advertised in aviation magazines.

Fig. 7-4. Interior reflective heatshield.

Fig. 7-5. Exterior windshield cover.

EXTERIOR CARE

Complete washing with automotive-type cleaners will produce acceptable results, and the materials used will be cheaper than "aircraft cleaners." Automotive wax will also provide adequate protection for painted or unpainted surfaces. When I say wax, that includes the new "space-age" silicone preparations called sealers.

An inexpensive product on the market for washing airplanes, cars, and boats, is Protect All Quick and Easy Wash that is applied by sponge or cloth to an exterior surface and wiped off with a chamois or dry cloth. It is ideal for washing airplanes where no running water is available. Another product to clean airplane exteriors is WashWax 777 and 787. Each is designed for a specific job: 777 for cleaning upper surfaces and 787 for lower, dirtier surfaces. WashWax is available from:

Aero Cosmetics
P.O. Box 460025
San Antonio, TX 78246
(800) 727-0747

Polish a bare metal airplane with the Cyclo Wonder Tool, a dual-orbital polishing tool. Personal experience has shown that this machine is well worth the investment if you want to keep a shiny unpainted airplane in tip-top condition. It is also useful when stripping before painting and will polish hard waxes (Fig. 7-6). For further information, contact:

Cyclo Manufacturing Co.
3841 Eudora Way, P.O. Box 2038
Denver, CO 80201

Fig. 7-6. Polishing tool.

Cyclo Manufacturing Co.

CLEANING SUPPLIES

Cleaning products in this list have been selected as representative of the market, in some cases based upon the manufacturer's advertised claims.

Micro-mesh is a specialized product designed to clean and polish Plexiglas—recommended for Plexiglas windshield maintenance. Plexiglas is a product of Rohm & Haas Company and is commonly used in aviation. Micro-mesh is available from:

Univair Aircraft Corporation
2500 Himalaya Road
Aurora, CO 80011
(303) 375-8882

Many items are available at most grocery stores or automotive supply houses. Pledge can be used on most hard surfaces including the windshield and vinyl surfaces, such as seats, dash panel, and doors. 409 is a heavy-duty cleaner for the hard-to-remove dirt and grime; keep 409 away from the windshield, instruments, and painted surfaces because the chemicals in 409 will mar the surface. Windex is excellent for small cleanup jobs, but never use Windex on the windshield or other aircraft windows, because Windex causes clouding and eventually the window would have to be replaced.

Armor-All is good for a final coat on the dash panel, kick panels, and the like. Gunk will degrease the engine area and front strut, then washes off with water. (When using Gunk in the engine compartment, keep the magnetos and alternator covered with plastic bags until the compartment is dry to prevent contamination.) Gunk is also good for cleaning the belly and is reasonably safe on painted surfaces if rinsed per instructions.

WD-40 is a general lubricant used to stop squeaks and reduce friction. Use WD-40 on cables, controls, seat runners, and doors—again, keep this product off the windows. (Refer to the owner's manual or a service manual for any manufacturer cautions regarding specific areas of lubrication.) LPS is a line of several lubricants available in various viscosities, similar to WD-40. Rubbing compound is used to clean away stains from exhaust; use rubbing compound carefully because paint might be removed with the stains.

Touch-up paint specifically marketed for airplanes is hard or impossible to find. A well-stocked automotive supply house will have inexpensive cans of spray paint or very small bottles of paint to match almost any automotive color and probably any airplane color. Try to find one that closely matches your airplane's base coat of paint or trim paint. Remember that a touch-up is just that, not a complete repaint, and don't expect it to be more. Touching-up small blemishes, chips, and the like, protects the airframe and perhaps makes the appearance slightly better.

8

Preventive maintenance

AN OWNER CAN DO MANY THINGS to keep an airplane properly maintained. Mechanical work can be a part of owner maintenance, but this is not to say a licensed mechanic is unnecessary because complex and sensitive components in every aircraft require expertise and an FAA license. Properly performed preventive maintenance gives the owner a better understanding of the airplane, affords substantial maintenance savings, and gives a feeling of accomplishment.

IMPORTANT ADVICE

The FARs require that all preventive maintenance work must be done in such a manner, and by use of materials of such quality, that the airframe, engine propeller, or assembly worked on will be at least equal to its original condition.

I strongly advise that before you undertake any of these allowable preventive maintenance procedures, you discuss your plans with a licensed mechanic. The instructions and advice you receive might help you avoid making costly mistakes. If the mechanic charges you for the consultation, it will be money well-spent.

TOOLS

Get your own tools. Don't borrow from your friendly mechanic, otherwise the mechanic won't be friendly very long. A small quantity of quality tools should allow the owner to perform preventive maintenance. A carrying bag or box will be very handy to keep these tool recommendations in order and protected (Supplement these with any other tools that you use repeatedly.):

- Multipurpose knife (Swiss Army knife)
- ⅜-inch ratchet drive (flexible head optional)
- 2, 4, 6-inch ⅜-inch extensions
- Sockets from ⅜-inch to ¾-inch (in 1/16-inch increments)
- 6-inch crescent wrench
- 10-inch monkey wrench

- 6- or 12-point closed (box) wrenches from ⅜-inch to ¾-inch
- Set of open-end wrenches from ⅜-inch to ¾-inch
- Slip-joint pliers (medium size)
- Phillips screwdriver set
- Blade screwdriver set
- Plastic electrical tape
- Container of aircraft-grade assorted nuts and bolts and screws
- Set of spark plugs
- 10x magnifying glass
- Inspection mirror

THE FAA SAYS

Federal Aviation Regulations specify that preventive maintenance may be performed by the owner of an airplane not utilized in commercial service. In case your FBO is concerned about your attempts at saving money, here is a partial reprint of Advisory Circular 150/5190-2A (April 4, 1972):

> d. Restrictions on Self-Service: Any unreasonable restriction imposed on the owners and operators of aircraft regarding the servicing of their own aircraft and equipment may be considered as a violation of agency policy. The owner of an aircraft should be permitted to fuel, wash, repair, paint, and otherwise take care of his own aircraft, provided there is no attempt to perform such services for others. Restrictions which have the effect of diverting activity of this type to a commercial enterprise amount to an exclusive right contrary to law.

FAA is allowing the owner of an aircraft to save hard-earned dollars—and to become very familiar with the airplane, which no doubt contributes to safety.

PROPER PARTS

All parts on an airplane must come from the original manufacturer or have FAA-PMA (Federal Aviation Administration-Parts Manufacture Approval). The Cessna 172 has many parts on it that are identical to parts available from the automotive industry: alternators, gauges, vee-belts, and the like. The automotive industry is Cessna's source for these parts; however, this does not mean you can just go to the local car parts store and purchase replacements for use in your airplane.

Unless the part is from Cessna as an original equipment manufacturer (OEM) part or is otherwise FAA-PMAd, you cannot legally use that part on an airplane. Ramifications include an unairworthy airplane that is not legal to fly and has no insurance coverage as a result of the bogus part.

Sadly, the identical parts (airplane and automotive) might vary in price by as much as 500–600 percent. For this additional cost, you will receive a part that has been thoroughly inspected and found to be airworthy.

LOGBOOK REQUIREMENTS

Entries must be made in the appropriate logbook whenever preventive maintenance is performed. An aircraft cannot be legally flown without proper notations regarding the maintenance. A logbook entry must include:

- Description of work done
- Date work is completed
- Name of the person doing the work
- Signature and certificate number of the person performing the preventive maintenance

PREVENTIVE MAINTENANCE

According to the FAA, "simple or minor preservation operations and the replacement of small standard parts not involving complex assembly operations" constitutes preventive maintenance. FARs list 28 recognized preventive maintenance items in Appendix A of Part 43; items applicable to the Model 172 airplanes are included here. Procedural instructions follow the list.

Preventive maintenance items

1. Removal, installation and repair of landing gear tires.
2. (Does not apply to the 172.)
3. Servicing landing gear struts by adding oil, air, or both.
4. Servicing landing gear wheel bearings, such as cleaning and greasing.
5. Replacing defective safety wiring or cotter pins.
6. Lubrication not requiring disassembly other than removal of nonstructural items such as cover plates, cowlings, and fairings.
7. (Does not apply to the 172.)
8. Replenishing hydraulic fluid in the hydraulic reservoir.
9. Refinishing decorative coatings of the fuselage, wing, and tail-group surfaces (excluding balanced control surfaces), fairings, cowlings, landing gear, cabin, or cockpit interior when removal or disassembly of any primary structure or operating system is not required.
10. Applying preservative or protective material to components when no disassembly of any primary structure or operating system is involved and when such coating is not prohibited or is not contrary to good practices.
11. Repairing upholstery and decorative furnishings of the cabin or cockpit when it does not require disassembly of any primary structure or operating system or affect the primary structure of the aircraft.
12. Making small, simple repairs to fairings, nonstructural cover plates, cowlings, and small patches; also, making reinforcements not changing the contour so as to interfere with the proper airflow.

13. Replacing side windows where that work does not interfere with the structure or any operating system, such as controls and electrical equipment.
14. Replacing safety belts.
15. Replacing seats or seat parts with replacement parts approved for the aircraft, not involving disassembly of any primary structure or operating system.
16. Troubleshooting and repairing broken landing light wiring circuits.
17. Replacing bulbs, reflectors, and lenses of position and landing lights.
18. Replacing wheels and skis where no weight and balance computation is required.
19. Replacing any cowling not requiring removal of the propeller or disconnection of flight controls.
20. Replacing or cleaning spark plugs and setting of spark plug gap clearance. (For an excellent package of information about spark plugs, write to any of the plug manufacturers. You'll receive usage charts, how-tos, and a color chart to grade used plugs, which helps identify potential problems.)
21. Replacing any hose connection except hydraulic connections.
22. Replacing prefabricated fuel lines.
23. Cleaning fuel and oil strainers.
24. Replacing batteries and checking fluid level and specific gravity.
25. (Does not apply to the 172.)
26. (Does not apply to the 172.)
27. Replacement or adjustment of nonstructural standard fasteners incidental to operations.
28. (Does not apply to the 172.)

PROCEDURAL INSTRUCTIONS

The following instructions and notes will help you perform preventive maintenance and understand other maintenance. You should have a service manual for the 172; if you do not have a manual and wish to purchase one, contact:

Essco
1615 S. Arlington St.
Akron, OH 44306
(216) 724-1249

Another excellent source of information for the airplane owner is FAA AC 20-106, *Aircraft Inspection for the General Aviation Aircraft Owner*, which is available from:

Superintendent of Documents
U.S. Government Printing Office
Washington, DC 20402

When contacting the Government Printing Office, ask for the *Subject Bibliography on Aviation Information and Training Materials*. It is free and gives titles and short blurbs about many available publications the airplane owner or pilot might want to purchase.

MAIN WHEELS

Main wheel removal

Note: It is not necessary to remove the main wheel to reline brakes or remove brake parts, other than the brake disc or torque plate. Refer to Figs. 8-1 through 8-4, which illustrate wheel maintenance.

1. Jack up one main wheel of the aircraft at a time. Place the jack at the universal jack point; do not use the brake casting as a jacking point. When using the universal jack point, flexibility of the gear strut will cause the main wheel to slide inboard as the wheel is raised, tilting the jack. The jack must be lowered for a second operation. Jacking both main wheels simultaneously with universal jack points is not recommended.
2. Remove the speed fairing (if installed).
3. Remove the hub cap, cotter pin, and axle nut.
4. Remove the bolts and washers attaching back plate to brake cylinder and remove back plate.
5. Pull the wheel off the axle.

Main wheel disassembly

1. Remove the valve core and deflate the tire. Break the tire beads loose from the wheel rims. **Warning:** Injury can result from attempting to separate wheel halves with the tire inflated. Avoid damaging wheel flanges when breaking tire beads loose.
2. Remove the through-bolts and separate the wheel halves, removing the tire, tube, and brake disc.
3. Remove the grease seal rings, felts, and bearing cones from the wheel halves.

Main wheel inspection and repair

1. Clean all metal parts and the grease seal felts in solvent and dry thoroughly.
2. Inspect the wheel halves for cracks. Cracked wheel halves shall be discarded and new parts used. Sand out nicks, gouges, and corroded areas. When the protective coating has been removed, the area should be cleaned thoroughly, primed with zinc chromate, and repainted with aluminum lacquer.
3. If excessively warped or scored, the brake disc should be replaced with a new part. Sand smooth small nicks and scratches.
4. Carefully inspect the bearing cones and cups for damage and discoloration. After cleaning, pack the cones with clean aircraft wheel bearing grease before installing them in the wheel half.

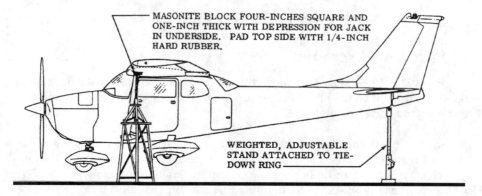

MASONITE BLOCK FOUR-INCHES SQUARE AND ONE-INCH THICK WITH DEPRESSION FOR JACK IN UNDERSIDE. PAD TOP SIDE WITH 1/4-INCH HARD RUBBER.

WEIGHTED, ADJUSTABLE STAND ATTACHED TO TIE-DOWN RING

UNIVERSAL JACK POINT (PART NO. 10004-98) AVAILABLE FROM THE CESSNA SERVICE PARTS CENTER

NOTE

Wing jacks available from the Cessna Service Parts Center are REGENT Model 4939-30 for use with the SE-576 wing stands. Combination jacks are the REGENT Model 4939-70 for use without wing stands. The 4939-70 jack (70-inch) may be converted to the 4939-30 jack (30-inch) by removing the leg extensions and replacing lower braces with shorter ones. The base of the adjustable tail stand (SE-767) is to be filled with concrete for additional weight as a safety factor. The SE-576 wing stand will also accommodate the SANCOR Model 00226-150 jack. Other equivalent jacks, tail stands, and adapter stands may be used.

1. Lower aircraft tail so that wing jack can be placed under front wing spar just outboard of wing strut.

2. Raise aircraft tail and attach tail stand to tie-down ring. BE SURE the tail stand weighs enough to keep the tail down under all conditions and is strong enough to support aircraft weight.

3. Raise jacks evenly until desired height is reached.

4. The universal jack point may be used to raise only one main wheel. Do not use brake casting as a jack point.

CAUTION

When using the universal jack point, flexibility of the gear strut will cause the main wheel to slide inboard as the wheel is raised, tilting the jack. The jack must be lowered for a second operation. Jacking both main wheels simultaneously with universal jack points is not recommended.

Fig. 8-1. Proper jacking. Cessna Aircraft Company

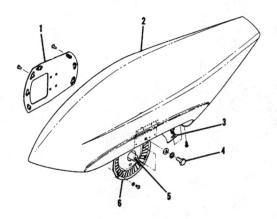

	Attach Plate	3.	Scraper	5.	Axle Nut
	Speed Fairing	4.	Bolt	6.	Hub Cap

Fig. 8-2. Main wheel speed fairing. Cessna Aircraft Company

Main wheel assembly

1. Insert the through-bolts through the brake disc and position it in the inner wheel half, using the bolts to guide the disc. Ascertain that the disc is bottomed in the wheel half.
2. Position the tire and tube with the tube inflation valve through the hole in the outboard wheel half.
3. Place the inner wheel half in position on the outboard wheel half. Apply a gentle force to bring the wheel halves together. While maintaining the force, assemble a washer and nut on one through-bolt and tighten snugly. Assemble the remaining washers and nuts on the through-bolts and torque to the value marked on the wheel. **Caution:** Uneven or improper torque of the through-bolt nuts can cause failure of the bolts, with resultant wheel failure.
4. Clean and pack the bearing cones with new aircraft wheelbearing grease.
5. Assemble the bearing cones, grease seal belts, and rings into the wheel halves.
6. Inflate the tire to seat the tire beads, then adjust inflation to the correct pressure prescribed in the owner's manual.

Main wheel installation

1. Place the wheel assembly on the axle.
2. Install the axle nut and tighten until a slight bearing drag is obvious when the wheel is rotated. Back off the axle nut to the nearest castellation and install a cotter pin.

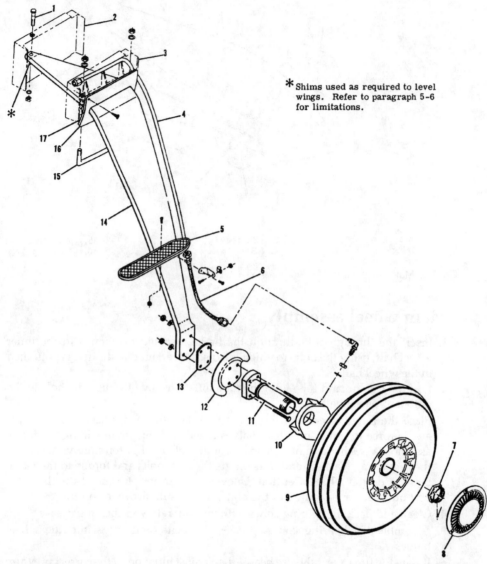

* Shims used as required to level
wings. Refer to paragraph 5-6
for limitations.

1.	Bolt				
2.	Inboard Forging	8.	Hub Cap		
3.	Outboard Forging	9.	Wheel Assembly		
4.	Brake Line	10.	Brake Cylinder	14.	Spring Strut
5.	Step	11.	Axle	15.	U-Bolt
6.	Brake Hose	12.	Brake Disc Cover Plate	16.	Fuselage Fairing
7.	Wheel Nut	13.	Shim	17.	Seal

Fig. 8-3. Main gear installation. Cessna Aircraft Company

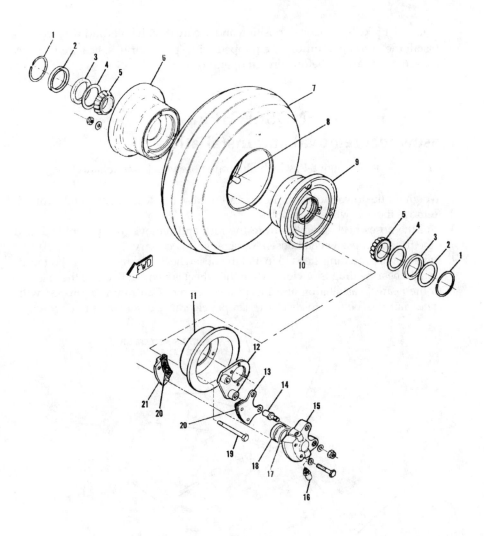

1.	Snap Ring	8.	Tube
2.	Grease Seal Ring	9.	Inner Wheel Half
3.	Grease Seal Felt	10.	Bearing Cup
4.	Grease Seal Ring	11.	Brake Disc
5.	Bearing Cone	12.	Torque Plate
6.	Outer Wheel Half	13.	Pressure Plate
7.	Tire	14.	Anchor Bolt

15.	Brake Cylinder
16.	Brake Bleeder
17.	O-Ring
18.	Piston
19.	Thru-Bolt
20.	Brake Lining
21.	Back Plate

Fig. 8-4. Wheel and brake assembly. Cessna Aircraft Company

Main wheels 151

3. Place the brake back plate in position and secure it with bolts and washers.
4. Install the hub cap. **Caution:** If equipped with speed fairings, be sure to check the scraper clearance before aircraft operation.

NOSEWHEEL
Nosewheel removal and installation

Refer to Figs. 8-5 through 8-7, which illustrate nosewheel maintenance.

1. Weight or tie down the tail of the aircraft to raise the nosewheel off the ground.
2. Remove the nosewheel axle bolt.
3. Pull the nosewheel assembly from the fork and remove the spacers and axle tube from the nosewheel. Loosen the scraper if necessary.
4. Reverse the preceding steps to install the nosewheel. Tighten the axle bolt until a slight bearing drag is obvious when the wheel is rotated. Back off the axle nut to the nearest castellation and install a cotter pin. **Caution:** If equipped with speed fairings, be sure to check the scraper clearance before aircraft operation.

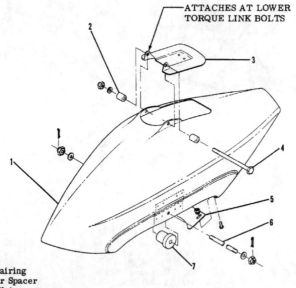

1. Speed Fairing
2. Tow-Bar Spacer
3. Cover Plate
4. Fork Bolt

5. Scraper
6. Axle Stud
7. Ferrule

Fig. 8-5. Nosewheel speed fairing. Cessna Aircraft Company

Nosewheel disassembly

1. Remove the hub cap, completely deflate the tire, and break the tire beads loose. **Warning:** Injury can result from attempting to separate wheel halves with the tire inflated. Avoid damaging wheel flanges when breaking tire beads loose.

2. Remove through-bolts and separate wheel halves.
3. Remove the tire and tube from the wheel halves.
4. Remove the bearing retaining rings, grease felt seals, and bearing cones.

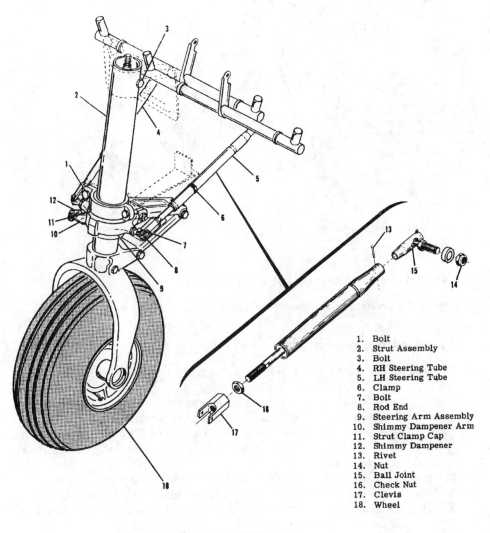

1. Bolt
2. Strut Assembly
3. Bolt
4. RH Steering Tube
5. LH Steering Tube
6. Clamp
7. Bolt
8. Rod End
9. Steering Arm Assembly
10. Shimmy Dampener Arm
11. Strut Clamp Cap
12. Shimmy Dampener
13. Rivet
14. Nut
15. Ball Joint
16. Check Nut
17. Clevis
18. Wheel

CAUTION

When installing cap (11), check the gap between the cap and
the strut fitting before the attaching bolts are tightened. Gap
tolerance is .010" minimum and .016" maximum. If gap ex-
ceeds maximum tolerance, install shims, Part No. 0543042-1
(.016") and Part No. 0543042-2 (.032"), as required to obtain
gap tolerance. Replace the cap if gap is less than minimum,
again using the shims to obtain proper gap. Install shims as
equally as possible between sides.

Fig. 8-6. Nose gear installation. Cessna Aircraft Company

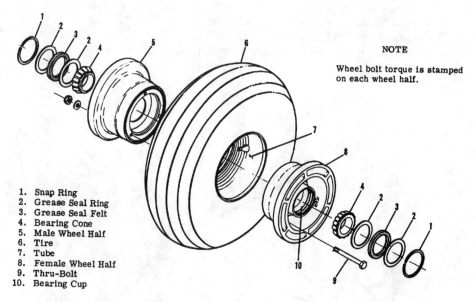

1. Snap Ring
2. Grease Seal Ring
3. Grease Seal Felt
4. Bearing Cone
5. Male Wheel Half
6. Tire
7. Tube
8. Female Wheel Half
9. Thru-Bolt
10. Bearing Cup

Fig. 8-7. Nosewheel.Cessna Aircraft Company

Nosewheel inspection and repair

Follow the procedures for main wheel inspection and repair.

Nosewheel assembly

1. Insert the tire and tube on a wheel half and position the valve stem through the hole in the wheel half.
2. Insert the through-bolts, position the other wheel half, and secure with nuts and washers. Take care to avoid pinching the tube between the wheel halves. Tighten the bolts evenly to the torque value marked on the wheel. **Caution:** Uneven or improper torque of the through-bolt nuts can cause failure of the bolts, with the resultant wheel failure.
3. Clean and pack the bearing cones with new aircraft wheel bearing grease.
4. Assemble the bearing cones, seals, and retainers into the wheel halves.
5. Inflate the tire to seat the tire beads, then adjust to the correct pressure prescribed in the owner's manual.
6. Install the spacers, axle tube, and hub caps, and install wheel assembly as described in nosewheel removal and installation.

Nose gear shock strut

Refer to Figs. 8-8 through 8-11, which illustrate shock strut maintenance. The shock strut requires periodic examination to ensure a proper level of hydraulic fluid and proper air pressure inflation. To service the nose gear strut, proceed as follows:

1. Remove the valve cap and release the air pressure.
2. Remove the valve housing.
3. Compress the nose gear to its shortest length and fill the strut with hydraulic fluid to the bottom of the filler hole.
4. Raise the nose of the aircraft, extend and compress the strut several times to expel any entrapped air, then lower the nose of the aircraft and repeat step 3.
5. With the strut compressed, install the valve housing assembly.
6. With the nosewheel off the ground, inflate the strut (refer to the owner's manual for exact pressure). **Note:** Keep the nose gear shock strut clean of dust and grit that might damage the seals in the strut barrel.

Fig. 8-8. Nose gear strut needing attention.

LUBRICATION

Refer to Figs. 8-12 through 8-14, which illustrate aircraft lubrication requirements. Before adding grease to grease fittings, wipe dirt from the fitting. Lubricate until grease appears around the parts being lubricated, then wipe excess grease away.

Wheel bearings should be cleaned and repacked at 500-hour intervals, unless heavy-use or dirt-strip operations occur; if the latter is the case, then service the wheel bearings every 100 hours.

Lubricate the nose gear torque links every 50 hours of operation (more often under dusty conditions).

Engine lubrication will be detailed in the aircraft owner's manual; however, as a rule of thumb, never exceed 50 hours between oil changes. Oil is cheap; when has an engine failed due to clean oil?

Oil changing for the airplane owner is no more complicated than for the family automobile, except have the airplane's oil analyzed. Engine oil analysis indicates what part of the engine is wearing and, over a period of time, how quickly that part is wear-

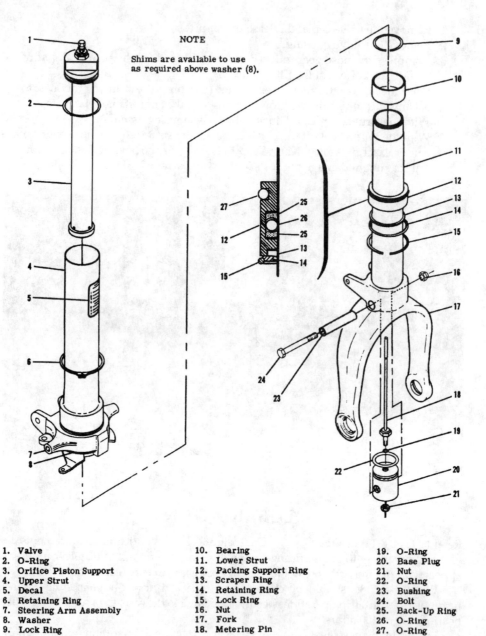

NOTE

Shims are available to use
as required above washer (8).

1. Valve	10. Bearing	19. O-Ring
2. O-Ring	11. Lower Strut	20. Base Plug
3. Orifice Piston Support	12. Packing Support Ring	21. Nut
4. Upper Strut	13. Scraper Ring	22. O-Ring
5. Decal	14. Retaining Ring	23. Bushing
6. Retaining Ring	15. Lock Ring	24. Bolt
7. Steering Arm Assembly	16. Nut	25. Back-Up Ring
8. Washer	17. Fork	26. O-Ring
9. Lock Ring	18. Metering Pin	27. O-Ring

Fig. 8-9. Nosewheel gear strut. Cessna Aircraft Company

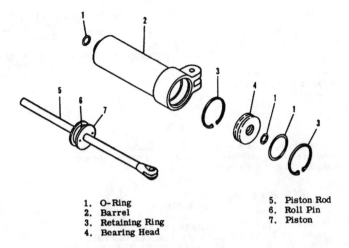

1. O-Ring
2. Barrel
3. Retaining Ring
4. Bearing Head

5. Piston Rod
6. Roll Pin
7. Piston

Fig. 8-10. Nose gear shimmy damper. Cessna Aircraft Company

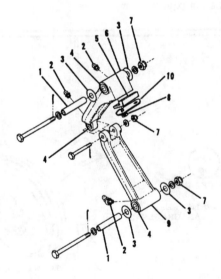

NOTE

Tighten bolts (8) to 20-25 pound-inches, then safety the bolts by bending tips of safety lug (10).

Tighten nuts (7) snugly, then tighten to align next castellation with cotter pin hole.

Shims (3) are available to use as required to remove any looseness.

1. Spacer
2. Grease Fitting
3. Shim

4. Bushing
5. Stop Lug
6. Upper Torque Link
7. Nut

8. Bolt
9. Lower Torque Link
10. Safety Lug

Fig. 8-11. Nose gear torque links. Cessna Aircraft Company

FREQUENCY (HOURS)

(50)   [100]   ◇ 500 ◇

┌─────────────────────────────┐
│ WHERE NO INTERVAL IS SPECIFIED, │
│ LUBRICATE AS REQUIRED AND   │
│ WHEN ASSEMBLED OR INSTALLED.   │
└─────────────────────────────┘

METHOD OF APPLICATION

HAND   GREASE   OIL   SYRINGE
        GUN   CAN   (FOR POWDERED
                         GRAPHITE)

NOTE

The military specifications listed are not mandatory, but are intended as
guides in choosing satisfactory materials. Products of most reputable
manufactures meet or exceed these specifications.

LUBRICANTS

PG —— MIL-G-6711   POWDERED GRAPHITE
GG—— MIL-G-7711   GENERAL PURPOSE GREASE
GA —— MIL-G-25760   AIRCRAFT WHEEL BEARING GREASE
GH —— MIL-G-23827   AIRCRAFT AND INSTRUMENT GREASE
GL —— MIL-G-21164   HIGH AND LOW TEMPERATURE GREASE
OG—— MIL-G-7870   GENERAL PURPOSE OIL
PL —— VV-P-236   PETROLATUM
GS ——   SIL-GLYDE (OR EQUIVALENT)

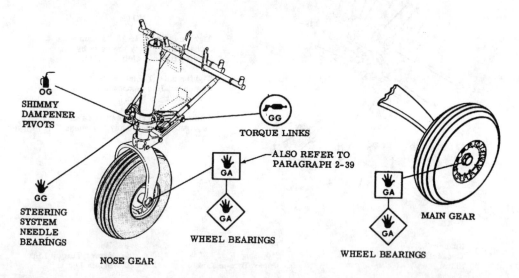

OG
SHIMMY
DAMPENER
PIVOTS

GG
TORQUE LINKS

ALSO REFER TO
PARAGRAPH 2-39
GA

GG
STEERING
SYSTEM
NEEDLE
BEARINGS

GA
WHEEL BEARINGS

NOSE GEAR

GA
MAIN GEAR

GA
WHEEL BEARINGS

Fig. 8-12. Lubrication chart. Cessna Aircraft Company

ing. Engine oil analysis services are available from several companies that advertise in
Trade-A-Plane and other aviation journals.

At oil change time, you might decide to put an oil additive in the crankcase, along
with the regular engine oil. The object of these additives is to reduce engine wear, resulting in longer engine life. (**Note:** Lycoming and Continental state that the use of oil

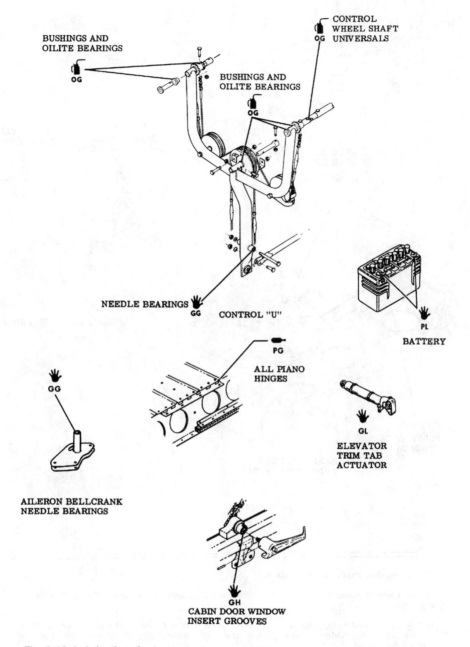

BUSHINGS AND
OILITE BEARINGS

OG

CONTROL
WHEEL SHAFT
UNIVERSALS
OG

BUSHINGS AND
OILITE BEARINGS

OG

NEEDLE BEARINGS
GG

CONTROL "U"

PL

BATTERY

PG

ALL PIANO
HINGES

GG

GL

ELEVATOR
TRIM TAB
ACTUATOR

AILERON BELLCRANK
NEEDLE BEARINGS

GH
CABIN DOOR WINDOW
INSERT GROOVES

Fig. 8-13. Lubrication chart. Cessna Aircraft Company

additives will void the manufacturer warranty.) One additive is Microlon, a Teflon product, approved by the FAA under FAR 33. Microlon is added to the oil only once in the life of the engine. The manufacturer claims that after introduction to the crankcase oil, the product bonds to all surfaces within the engine, providing extremely good lubricating properties. Microlon is available from:

Microlon, Inc.
1305 Fraser Street Suite D-5
Bellingham, WA 98226
(800) 962-4152
(206) 647-1350

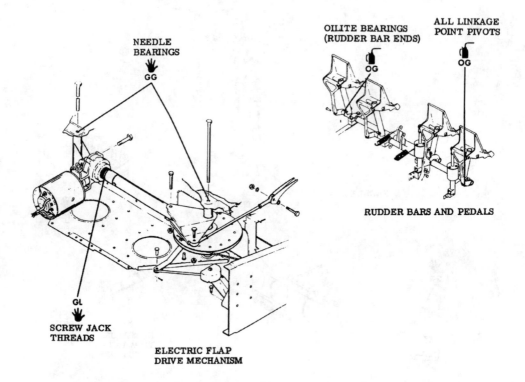

NEEDLE BEARINGS
GG

OILITE BEARINGS (RUDDER BAR ENDS)
OG

ALL LINKAGE POINT PIVOTS
OG

RUDDER BARS AND PEDALS

GL
SCREW JACK THREADS

ELECTRIC FLAP DRIVE MECHANISM

═══════════════════ NOTES ═══════════════════

Sealed bearings require no lubrication.

Do not lubricate roller chains or cables except under seacoast conditions. Wipe with a clean, dry cloth.

Lubricate unsealed pulley bearings, rod ends, Oilite bearings, pivot and hinge points, and any other friction point obviously needing lubrication, with general purpose oil every 1000 hours or oftener if required.

Paraffin wax rubbed on seat rails will ease sliding the seats fore and aft.

Lubricate door latching mechanism with MIL-F-7711 general purpose grease, applied sparingly to friction points, every 1000 hours or oftener if binding occurs. No lubrication is recommended on the rotary clutch.

Fig. 8-14. Lubrication chart. Cessna Aircraft Company

HYDRAULIC BRAKE SYSTEMS

Check brake master cylinders and refill with hydraulic fluid as required every 100 hours.

MINOR FAIRING SURFACE REPAIRS

Fairing repair for fiberglass

1. Remove the fairing and drill a stop hole at the end of a crack to prevent further extension of the crack.
2. Dust the inside of the surface to be repaired with baking soda, then position a small piece of fiberglass cloth over the crack and saturate the entire area with cyanoacrylate (super glue). Repeat step 2.
3. From the outside, fill the crack with baking soda and harden with cyanoacrylate.
4. Sand smooth and repaint.

Polyfix is a commercially available product for repairing most plastic materials, available from:

Redam Corp.
321 Ross Ave.
Hamilton, OH 45013
(800) 234-4583
(513) 863-9393

Fairing repair for metal

1. Remove the fairing and drill a stop hole at the end of a crack to prevent further extension of the crack.
2. Obtain a separate piece of metal that matches the fairing metal. Cut off a portion of the separate piece that is approximately 1 inch wider than each side of the crack.
3. Drill matching holes through the patch and fairing, then rivet together.
4. Touch up with paint as needed.

PAINT TOUCH-UP

Touching up small areas on the wings or fuselage of an airplane is very easy and makes a marked improvement in appearance.

1. Thoroughly wash the area to be touched up. All preservatives such as wax and silicone products must be removed.
2. Any loose or flaking paint must be removed. Carefully use very fine sandpaper for this purpose. Do not sand the bare metal.

3. If bare metal is exposed, it must be primed with an aircraft zinc chromate primer.
4. Using sweeping spray strokes, apply at least two coats of touch-up paint in a color matching the original surface.

Apply at least two coats of touch-up paint in a color matching the original surface. Sweep the area while spraying to avoid runny paint and air bubbles.

CABIN UPHOLSTERY

Probably nothing is more difficult for the do-it-yourselfer than the removal and installation of an airplane's headliner (Fig. 8-15). These instructions will improve the chances of finishing the job with the fewest hassles, producing a like-new look inside your 172.

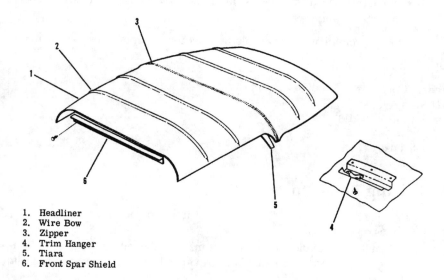

1. Headliner
2. Wire Bow
3. Zipper
4. Trim Hanger
5. Tiara
6. Front Spar Shield

Fig. 8-15. Cabin headliner. Cessna Aircraft Company

Headliner removal

1. Remove the sun visors, all inside finish strips and plates, door post upper shields, front spar trim shield, dome light panel, and any other visible retainers securing the headliner. (Save any reusable screws and fastening hardware in an organized manner, perhaps in an old plastic ice tray.)
2. Work the edges of the headliner free from the metal tabs that hold the fabric.
3. (**Note:** Always work from front to rear when removing the headliner; it is impossible to detach the wire bows when working from rear to front.) Starting at the front of the headliner, work the headliner down, removing the screws through the metal tabs that hold the wire bows to the cabin top. Pry loose the

outer ends of the bows from the retainers above the doors. Detach each wire bow in succession.

4. Remove the headliner assembly and bows from the airplane. **Note:** Due to the difference in length and contour of the wire bows, each bow should be tagged to assure proper location in the headliner.
5. Remove the soundproofing panels that are held in place with glue.

Headliner installation

1. Before installing the headliner, check all items concealed by the headliner to see that they are mounted securely. Use wide cloth tape to secure loose wires to the fuselage, and to seal any openings in the wing roots. Straighten any tabs bent during the removal of the old headliner.
2. Apply cement to the skin areas where the soundproofing panels are not supported by wire bows, and press the panels into place.
3. Insert wire bows into the headliner seams, and secure the rearmost edges of the headliner after positioning the two bows at the rear of the headliner. Stretch the material along the edges to make sure it is properly centered, but do not stretch it tight enough to destroy the ceiling contours or distort the wire bows. Secure the edges of the headliner with sharp tabs, or, where necessary, rubber cement.
4. Work the headliner forward, installing each wire bow in place with the tabs. Wedge the ends of the wire bows into the retainer strips. Stretch the headliner just taut enough to avoid wrinkles and maintain a smooth contour.
5. When all bows are in place and fabric edges are secured, trim off any excess fabric and reinstall all items removed.

Seat re-covery

Re-covering the seats is generally easier than installing a headliner (Figs. 8-16 and 8-17). **Important:** Work in a well-ventilated area or the glue will have you flying without the airplane. Ideally, contract the job with a professional aviation interior shop or contact a supplier of complete interiors or slip covers (Fig. 8-18). Slip covers are recommended if you are the hands-on type and want to save money. *Trade-A-Plane* has numerous advertisements for upholstery suppliers. An alternative source of seats and carpets is aircraft salvage companies. For more information contact:

Cooper Aviation Supply Co.
2149 E. Pratt Blvd.
Elk Grove Village, IL 60007
(708) 364-2600

Garrett Leather Corp.
P.O. Box 29
359 Niagara Street
Buffalo, NY 14201
(800) 342-7738
(716) 852-7720

Fig. 8-16. Standard seats found in a 1974 Skyhawk.

Fig. 8-17. Seat covers.

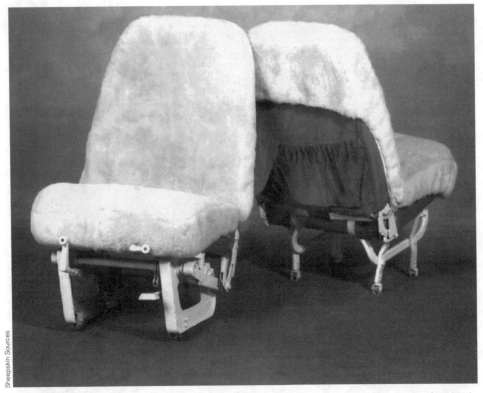

Fig. 8-18. Sheepskin seatcovers easily slip on over worn upholstery and are very comfortable in cool or cold weather.

SEATS AND SAFETY BELTS

Individual seats are equipped with manually operated reclining seat backs. Rollers permit the seats to slide forward and backward on seat rails. An adjustment mechanism inserts pins into holes along a metal track, which locks the seats in selected positions. Stops limit seat travel (Figs. 8-19 and 8-20).

Removal of a seat is accomplished by removing the stops and moving the seats forward and back on the rails to disengage them from the rails; installation is in reverse order. **Warning:** It is extremely important that the pilot's seat stops be installed because acceleration and deceleration might cause the seat to become disengaged from the seat rails and create a hazardous situation, especially during takeoff and landing.

Safety belts must be replaced when frayed or cut, or the latches become defective. Anchoring hardware should be replaced if faulty. Use only aviation-approved safety belt components.

SIDE WINDOW REPLACEMENT

A movable window, hinged at the top, is installed in the left door and might be installed in the right door. The window assembly can be replaced by pulling the hinge

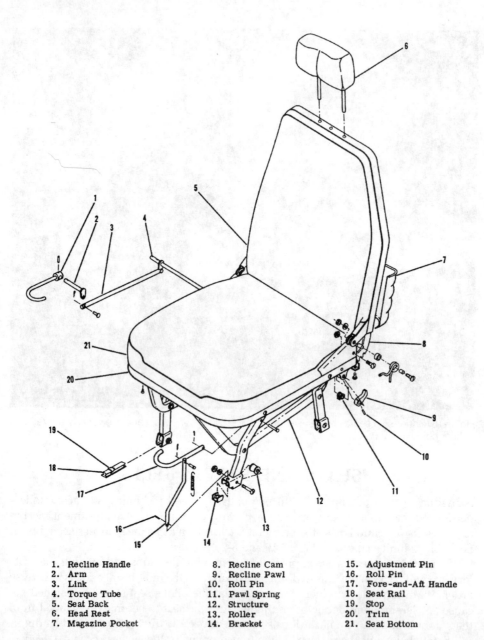

Fig. 8-19. Standard single seat. Cessna Aircraft Company

1. Recline Handle	8. Recline Cam	15. Adjustment Pin
2. Arm	9. Recline Pawl	16. Roll Pin
3. Link	10. Roll Pin	17. Fore-and-Aft Handle
4. Torque Tube	11. Pawl Spring	18. Seat Rail
5. Seat Back	12. Structure	19. Stop
6. Head Rest	13. Roller	20. Trim
7. Magazine Pocket	14. Bracket	21. Seat Bottom

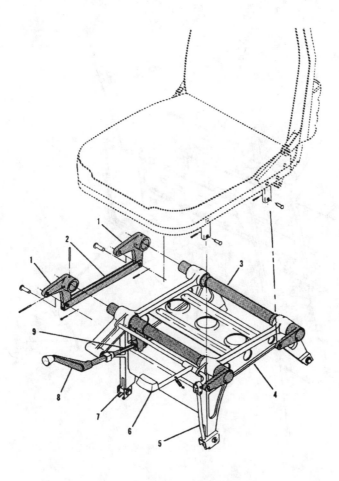

NOTE

Seat bottom, seat back, reclining
mechanism, and fore-and-aft
adjusting mechanism are similar
to the standard seat.

1. Bellcrank
2. Channel
3. Torque Tube

4. Seat Structure
5. Pin
6. Fore-and-Aft Adjustment Handle

7. Seat Roller
8. Vertical Adjustment Handle
9. Adjustment Screw

Fig. 8-20. Vertically adjustable seat. Cessna Aircraft Company

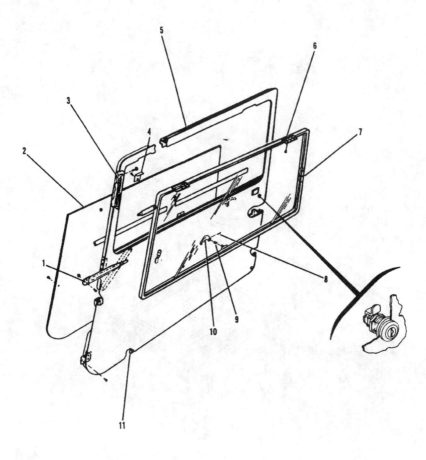

1.	Hinge	5.	Cabin Door		
2.	Upholstery Panel	6.	Window Hinge	9.	Spring
3.	Spring	7.	Frame Splice	10.	Latch
4.	Striker Plate	8.	Roll Pin	11.	Seal

Fig. 8-21. Cabin door and movable window. Cessna Aircraft Company

pins and disconnecting the window stop (Fig. 8-21). To remove the frame from the plastic, it is necessary to drill out the blind rivets where the frame is spliced. When replacing a window in a frame, make sure that the sealing strip and an adequate coating of a sealing compound (Presstite No. 579.6) are used all around the edges of the plastic panel.

BATTERY

Battery servicing involves adding distilled water to maintain the electrolyte level with the horizontal baffle plate at the bottom of the filler holes. Be sure to flush the area with plenty of clean water after refilling to wash away any spilled battery acid.

PARTS

It is sometimes very difficult to locate exact replacement parts for older airplanes. This is particularly true for structural parts and plastic items. Generally, hardware will be easier to locate due to its generic nature.

Cessna has a large parts supply system with a very large warehouse that fills orders within hours of receipt and usually ships the same day. Most Cessna 172 parts are in stock, except engines and complete wings. Engine and wing parts are available from other suppliers.

Cessna only sells parts to authorized Cessna dealers. This means an additional level of profit between the supplier and the user. Additionally, it is not unheard of for an authorized dealer to only sell parts with installation.

Be prepared to pay a premium for obscure parts from Cessna; however, for exhaust systems, windshields, and other items generally available from other sources, you will find Cessna prices to be competitive. Despite pricing issues, Cessna's parts can return an old 172 to original appearance and condition. For additional information, contact:

Cessna Aircraft Co.
Wichita, KS 67201
(316) 685-9111
(316) 941-6118 for service bulletin information
(800) 545-4611

Hardware and many other parts for the Cessna 172 are available from the following sources. Most will supply a catalog upon request:

Alexander Aeroplane Company, Inc.
900 S. Pine Hill Road
P.O. Box 909
Griffin, GA 30224
(800) 831-2949

Superflite
2149 E. Pratt Blvd.
Elk Grove Village, IL 60007
(800) 323-0611

Univair Aircraft Corp.
2500 Himalaya Road
Aurora, CO 80011
(303) 375-8882

Wag-Aero, Inc.
P.O. Box 181
1216 North Road
Lyons, WI 53148
(800) 766-1216

Parts are also advertised by many other suppliers in *Trade-A-Plane* and the *Aviation Telephone Directory*. The latter publication, available for purchase, provides telephone listings for aviation related businesses, airports, and FBOs, and has very good yellow pages of products and services. For further information contact:

Aviation Telephone Directory
515 West Lambert Road Suite D
Brea, CA 92621
(714) 990-5115

Alternate sources of plastic parts

Interior molded plastic parts such as door panels, post covers, access covers, headliners, seat trim backs, and the like, are available from:

Airflite Industries, Inc.
P.O. Box 8
Grand Ledge, MI 48837
(800) 345-7753
(517) 627-9322

Kinzie Industries, Inc.
P.O. Box 847
Alva, OK 73717
(405) 327-1565

Texas Aero Plastics
Northwest Regional Airport
Route 9 Box 17
Roanoke, TX 76262
(817) 491-4735

Stainless steel screw kits

Have you noticed all those rusted screws on an airplane? It's easy to replace them with nonrusting stainless steel screws. Kits containing screws in the proper number and of the proper size are the recommended way to purchase supplies for this job. Stainless steel screws on airplanes make sense because they don't rust and stain the surrounding area.

Be very careful that you don't strip out the screw holes when removing and installing screws. If you are replacing rusted or corroded screws, perhaps you should clean the screw hole for easier installation of the new screw.

Stainless steel screws are usually packaged as complete kits for specific airplanes by make, model, and year (Fig. 8-22). Each kit contains everything you will need. Kits are available from Sporty's, Air Components, Wil Neubert Aircraft Supply, J&M Aircraft, and from many FBOs. Direct supply is available from:

D&D Aircraft Supply
4 Stickney Terrace, P.O. Box 1200
Hampton, NH 03842
(800) 468-8000
(603) 926-8881

Trimcraft Aviation
P.O. Box 488
Genoa City, WI 53128
(414) 279-6896

Fig. 8-22. Trimcraft stainless steel screw kit for the 172.

9

Updating avionics

CESSNA 172S ARE VERY VERSATILE, providing reasonable speed, excellent reliability, and reasonable maintenance costs; however, one element that probably needs attention is the avionics, particularly on an old airplane or an airplane with minimal equipment (Figs. 9-1 through 9-3). General aviation has a comparative set of standards that can help determine which avionics should be installed in an airplane. FARs also set forth minimum avionic standards for certain operations, especially IFR, that are necessary. Consult the FARs for specific requirements. This table is one place to start:

Equipment	VFR	IFR
Transceiver	x	x
Navigational receivers	x	x
Distance measuring		x
Marker beacon receiver		x
Glideslope		x
Transponder	x	x
Audio panel		x
Emergency locator transmitter	x	x
Clock	x	x
Encoding altimeter		x

Area navigation, automatic direction finding, and loran equipment would also be nice to have in a panel. Anything extra added to the recommended minimums is all the better (Figs. 9-4 through 9-25).

AVIONICS ABBREVIATIONS

COMM or COM. VHF transceiver for voice radio communications.

NAV. VHF navigation receiver for utilizing VORs.

NAV/COMM or navcom. Combination of navigation receiver and transceiver in one unit.

LOC/GS or localizer/glideslope. Visual output is via a CDI for the localizer; a horizontal indicator shows aircraft position relative to the glideslope.

Fig. 9-1. Example of a very early Cessna 172 panel.

CDI. The course deviation indicator is panel-mounted and gives a visual output of the navigation receiver's data.

XPNDR or transponder. A radar beacon identifier that might have altitude encoding.

ADF. Automatic direction finder.

DME. Distance measuring equipment.

LORAN-C or loran. A very accurate radio and computer-based navigation system unrelated to VORs.

A-panel or audio panel. Controls radio equipment audio output.

ELT. Emergency locator transmitter (installation requirements according to FARs).

MBR. Marker beacon receiver.

FLYING REQUIREMENTS

Your flying determines your avionics needs. No doubt the pilot flying in heavy IFR conditions will require more equipment than the weekender going to the next airport for coffee. If you are a casual flier, operating on weekends or evenings, and do little cross-country flying, then you can get by with minimal equipment: navcom, transponder, and ELT.

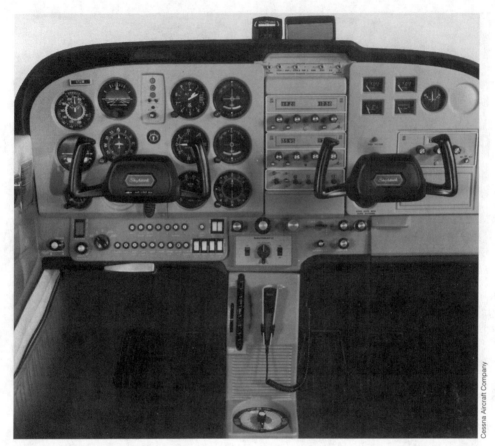

Fig. 9-2. 1974 Cessna 172M panel.

It wasn't too many years ago that cross-country flying was done by pilotage (reading charts and looking out the windows for checkpoints); today's aviator has become accustomed to the advantages of modern electronic black boxes; therefore, an airplane equipped in a minimal fashion would be inappropriate for the typical pilot of today.

If you enjoy VFR cross-country flying, as most family pilots do, you will need a little more equipment to ease your workload, and to give you backup in case of failure: dual navcom, DME, transponder, ELT, and loran.

The additional navcom can be used to lighten your workload, as well as provide backup in case of partial equipment failure. One more thing that will really ease your workload on long cross-country trips is autopilot; the unit doesn't have to be complex, a wing-leveler would be fine.

If you fly IFR—and the Cessna 172s are certainly capable of it—you will need still more equipment, in addition to the above: localizer and glideslope indicator, marker beacon receiver, ADF, altitude encoding transponder, and RNAV.

The entire IFR installation must be certified, so be prepared to spend some money if you are planning to completely reoutfit a plane for IFR. It might be worthwhile to consider a different airplane, rather than just avionics. Often you can purchase a newer

Fig. 9-3. 1977 R172 panel.

airplane, equipped as you want, for less than it would cost you to update your present airplane.

PANEL UPGRADES

Vacant spots in the instrument panel can be filled several ways; some are more expensive than others and some will provide better results than others.

New equipment

New equipment is state-of-the-art, offering the newest innovations, best reliability, and—best of all—a warranty. An additional benefit is the fact that the new solid-state electronics units draw considerably less electric power than did their tube-type predecessors. This is extremely important for the person wanting a full panel.

New avionics can be purchased from your local avionics dealer, or from a discount house. You can visit your local dealer and purchase all the equipment you want, and have it installed. Of course, this will be the most expensive route you can take when

Fig. 9-4. Val Com's transceiver. VAL Avionics Ltd.

Fig. 9-5. Narco's Star Nav is a fully digital navigation receiver with scanning, DME, and GS.

upgrading avionics; however, in the long run, this route might be the most cost-effective. You'll have new equipment, expert installation, and service backup. You will also have a nearby dealer for assistance if problems arise.

The discount house will be considerably cheaper for the initial purchase; however, you might be left out in the cold if warranty service is required. Certain manufacturers will not honor warranty service requests unless the equipment was purchased from and installed by an authorized dealer. Perhaps this sounds unfair to you; however, it

Fig. 9-6. Narco's MK12D state-of-the-art digital navcom with memory.

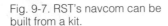

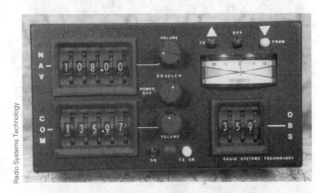

Fig. 9-7. RST's navcom can be built from a kit.

Fig. 9-8. King KX155 and KX165 navcom family.

will keep the authorized dealers in business. And if they stay in business, you can find them to repair your equipment.

Used equipment

Used avionics can be purchased from dealers or individuals. The aviation magazines and *Trade-A-Plane* are good sources of used equipment. Here are a few words of

Fig. 9-9. King KI203 VOR/LOC indicator.

Bendix King Radio Corporation

Fig. 9-10. RST's course deviation indicator for VOR, localizer, and glideslope.

Radio Systems Technology

caution about used avionics: Purchase nothing with tubes, nothing that is more than six years old, or nothing made by a defunct manufacturer, because tubes and other parts might be difficult to obtain. Also avoid any electronics sold "as is" or "working when removed."

Used equipment can be a wise investment, but it is very risky unless you happen to be an avionics technician, or have access to one. Do not purchase used avionics unless you are very familiar with the source. Further, do not purchase used avionics for primary IFR equipment.

Reconditioned equipment

Several companies advertise reconditioned avionics at bargain—or at least low—prices. The equipment has been removed from service and completely checked out by an avionics shop. Parts that have failed, are near failure, or are likely to fail will have

Fig. 9-11. Narco's Escort II navcom with digital VOR display is small enough to fit a standard 3-inch panel hole.

Narco Avionics

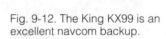

Fig. 9-12. The King KX99 is an excellent navcom backup.

Bendix King Radio Corporation

Fig. 9-13. Narco AT150 transponder.

Fig. 9-14. Collins TDR950 transponder.

Fig. 9-15. King KT76A transponder.

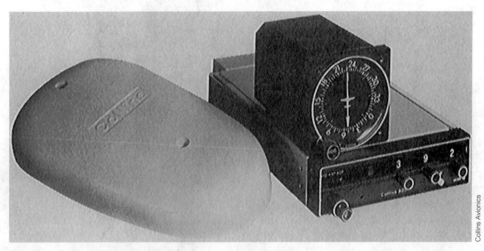

Fig. 9-16. Collins ADF650 and antenna.

Fig. 9-17. King KN64 distance measuring equipment.

Fig. 9-18. RST Long Ranger loran.

been replaced. These radios offer a fair buy for the airplane owner, and are usually warrantied by the seller; however, be advised that reconditioned is not new, although the original radio is probably only six or seven years old. Everything in the unit has been used, but not everything will be replaced during reconditioning. You will have some new parts and some old parts. Being aware of this drawback, I feel that reconditioned equipment purchases make sense for the budget-minded owner.

LORAN

Long range navigation, called loran or loran-C, is based upon low-frequency radio signals, rather than the VHF FAA navaids normally associated with flying. Actually, loran was not really intended for general aviation usage, but has become very popular. The current version of loran is C, which indicates computer-based.

Fig. 9-19. IIMorrow's Apollo Flybuddy loran uses Flybrary data cards to update a database of all public-use airports and VORs in the United States and Canada.

Fig. 9-20. IIMorrow's Flybrary data card being inserted into a Flybuddy.

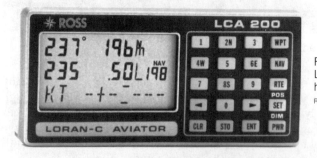

Fig. 9-21. Ross Engineering's LCA 200 loran Aviator has a high-visibility LCD display.

Ross Engineering Company

The newest versions of loran offer distinct advantages over normal VHF navaids such as VORs. Due to the propagation properties of radio waves at the frequencies utilized by loran, there is no VHF line-of-sight usable range limit. This means that unlike most VORs, usable only within a short range of 50 to 100 miles, loran is usable many hundreds of miles from the actual station. This opens up some very interesting possibilities for use.

Flying below 2000 feet agl in very isolated areas can be a limiting factor when navigating by use of standard VORs because the signals are line-of-sight, blocked by the

Fig. 9-22. The SportNav
hand-held loran is made in the
United States of America.

Voyager

Fig. 9-23. King KA134 audio
control panel.

Bendix King Radio Corporation

Fig. 9-24. Collins ARMR350
audio control panel with marker
beacon. Collins Avionics

Fig. 9-25. RST's audio panel
directs audio from various
sources as speakers or
headphones.

curvature and curvature and curvature and physical features of the earth at a low altitude. Loran is usable right on down to the ground.

Many loran units are reworked marine versions. (Loran transmitters are maintained by the U.S. Coast Guard, not the FAA.) They are not certified for IFR work; however, this does not mean they are incapable or inaccurate. This only means the manufacturer was unwilling to spend the many dollars necessary for IFR certification. It is also an indicator of price. The simpler uncertified versions are generally available for fewer than $1000 and IFR-certified units will cost more than $4000. Features vary from model to model and prices vary accordingly.

Without going into extensive theory about operation, the loran unit can, by simultaneously receiving and comparing several loran signals, determine exact location within a few feet, which is displayed on the readout as latitude and longitude. Waypoints that are typically programmed by the pilot can be used for navigation. The unit will then compare the known signals to the waypoint and can display course direction, elapsed time, estimated time enroute, distance traveled, distance to destination, and the like.

BOTTOM LINE RATIONALE

Save your money until you can purchase new equipment that has more features than previous models, plus smaller size and less weight, reduced electrical requirements, and better reliability. Additionally, due to inflation, current avionics are a better bargain than 20 years ago. Don't even think about trading in equipment that is currently working properly. You cannot replace it for what a dealer will give you. Keep it as your second system.

To some, the instrument panel is a functional device; to others it's a statement made by the owner. In either case, much care must be taken when filling up the panel—not just to provide instrumentation, but do it economically and functionally.

INTERCOMS AND HEADSETS

After flying for many continuous hours in an airplane, your ears will be ringing from all the constant noise; it is possible to damage your hearing with endless assaults of loud noise. These excerpts from Advisory Circular AC 91-35 examine noise considerations.

> Modern general aviation aircraft provide comfort, convenience, and excellent performance. At the same time that the manufacturers have developed more powerful engines, they have given the occupants better noise protection and control, so that today's aircraft are more powerful, yet quieter than ever. Still, the levels of sound associated with powered flight are high enough for general aviation pilots to be concerned about participating in continuous operations without some sort of personal hearing protection.
>
> Many long-time pilots have a mild loss of hearing. Many pilots report unusual amounts of fatigue after flights in particularly noisy aircraft. Many pilots have temporary losses of hearing sensitivity after flights, and many pilots have difficulty understanding transmissions from the ground, especially during critical periods under full power, such as takeoff.

Like carbon monoxide, noise exposure has harmful effects that are cumulative, (producing) a greater effect on the listener (as sound intensity increases and the length of exposure increases). A noise that could cause a mild hearing loss to a (pilot) who heard it once a week for a few minutes might make (that pilot) quite deaf if (lasting) for eight hours.

A remedy for pilots and passengers is a headphone intercom system (Figs. 9-26 through 9-28). Intercom systems come in all types and with varied capabilities. Some are an extension of the audio panel, primarily for the use of the pilot in his duties; others are completely portable.

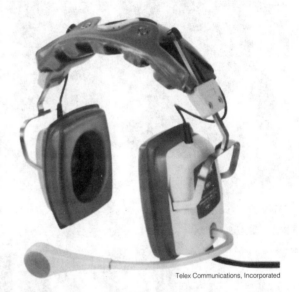

Fig. 9-26. Telex noise attenuating headsets.

Telex Communications, Incorporated

No matter what type you select, the ear protection will be controlled by the quality of the headphones. For proper ear protection, you must use full ear-cover headsets, not lightweight stereo headsets. Advertisements from several manufacturers are found in aviation periodicals and catalogs. Don't make a selection based solely on an advertisement. Talk to other pilots, then go to an aviation supply store and try a few. Pay particular attention to the weight because weight will become a fatigue factor during a long flight. After you find a system you like, purchase it and use it.

Just in passing, I have heard of some pilots who supply only headsets—no microphones—to passengers. I don't advocate this because flying is fun and conversation on a long flight without yelling might reduce fatigue for passengers.

DIGITAL AND NONDIGITAL DEVICES

Modern technology's digital readouts have attracted pilots that have added numerous bells and whistles to airplane panels. Among the more common are the digital outside air temperature, altitude, and electrical voltage gauges (Figs. 9-29 and 9- 30). CHT and EGT gauges, as noted in chapter 3, also have digital displays.

David Clark Company, Incorporated

Fig. 9-27. David Clark electronic noise-canceling headset and microphone system.

Radio Systems Technology

Fig. 9-28. RST intercom.

Heads-Up is a small stick-on card with a sensitized area that reacts by color change to the presence of deadly carbon monoxide in the cabin area. This simple device has no moving parts and might save your life (Fig. 9-31). Heads-Up is available from:

Sims Marketing
23 N. Gore, Suite 002
St. Louis, MO 63119-2300
(314) 961-0896

Fig. 9-29. Davtron digital OAT 301C. Davtron

Fig. 9-30. Davtron M655 indicates temperature (Fahrenheit or Celsius), pressure altitude, density altitude, and electrical system voltage. Davtron

Fig. 9-31. A carbon monoxide detector is an inexpensive safety item. Sims Marketing

VINTAGE AVIONICS

Aire-Sciences

Model	Navcom channels	Other
RT-551A	200/720	
RT-553	200/360	
RT-553A	200/720	
RT-557		Transponder
RT-667		Transponder
RT-777		Transponder
RT-787		Transponder
RT-887		Transponder
RT-563	200/360	
RT-563A	200/720	
RT-661A	200/720	
RT-773	200/360	

Cessna (ARC)

The Cessna 300 and 400 series navcom have come in 90-, 100-, 360-, and 720-channel versions. These units were first placed into production more than 30 years ago, and have been improved on a yearly basis. The 300/400 series also includes ADF and transponder equipment. The individual model numbers changed each year.

Genave

Model	Navcom channels	Other
Alpha 100	100	
Alpha 190	100/90	
Alpha 200	200/100	
Alpha 200A	200/100	
Alpha 200B	200/100	
Alpha 300	100/360	
Alpha 360	100/360	
Alpha 500	200/360	
Alpha 600	200/360	
Alpha 720	720	
Beta 500		Transponder
Beta 5000		Transponder
GA-1000	200/720	
Sigma 1500		ADF

King

Model	Navcom channels	Other
KR-80		ADF
KR-85		ADF
KR-86		ADF
KR-87		ADF
KT-75		Transponder
KT-76		Transponder
KT-78		Transponder
KX-100A	90/190	
KX-120	360	
KX-130	100/360	
KX-145	200/720	
KX-150A, B	100/100	
KX-155	200/720	
KX-160	100/360	
KX-165	200/720	
KX-170	200/360	
KX-170A	200/360	
KX-170B	200/720	
KX-175B	200/720	
KY-90A	90	
KY-95	360	
KY-195B	720	

Narco

Model	Navcom channels	Other
10A	200/360	
11A	360	
100	360	
110	200/360	
ADF-140		ADF
ADF-29		ADF
ADF30A		ADF
ADF-31(all)		ADF
AT-6		Transponder
AT-50		Transponder
AT-150		Transponder
Escort 110	100/110	
VHT-3	tune/19	
MK-2	tune/27	
MK-3	190/90	

Model	Navcom channels	Other
MK-4	tune/27	
MK-5	190	
MK-7	360	
MK-8	100	
MK-10	190/360	
MK-12(90)	100/90	
MK-12(360)	100/360	
MK-12A(90)	100/90	
MK-12A(360)	100/360	
MK-12B	100/360	
MK-12D	200/720	
MK-16	200/360	
MK-24	100/360	
Nav-11	200	
Nav-12	200	

Terra

Model	Navcom channels	Other
ML-200	200/100	
R250		Transponder
R360-200-1	200/360	
TX-720	720	

10

Aircraft painting

MOST OWNERS WOULD NEVER ATTEMPT to paint their airplane, but owners should know what a good paint job involves. Most of the information in this chapter is based upon material from Randolph Products Company, of Carlstadt, New Jersey. **Note:** If painting an aircraft becomes a do-it-yourself job, check the owner's manual or check with an inspection facility to determine whether or not the control surfaces of the aircraft require rebalancing as a result of painting.

STRIPPING ALUMINUM AIRCRAFT

Do not let paint remover come in contact with any fiberglass components of the aircraft such as radomes, wingtips, fairings, and the like. Make sure that these parts are well masked or removed from the aircraft while stripping is in progress.

While using remover, always wear rubber gloves and protect your eyes from splashes. If remover gets on skin, flush with plenty of water; if any comes in contact with your eyes, flood repeatedly with water and call a physician. Ensure adequate ventilation.

Rand O Strip remover (B-5000) is a fast-acting water wash paint remover designed for use on aircraft aluminum surfaces; it conforms to military specification MIL-R-25134.

Stir contents before proceeding. Apply liberally by brush or nonatomizing spray to metal surface. When brushing, be sure to brush only in one direction. Keep surface wet with remover. If an area dries before the paint film softens or wrinkles, apply more remover. It is sometimes advisable to lay an inexpensive polyethylene drop cloth over the applied remover in order to hold the solvents longer, giving more time for penetration of the film. After the paint softens and wrinkles, use a pressurized water hose to thoroughly flush off all residue.

In the case of an acrylic lacquer finish, the remover will only soften and will not wrinkle the film. A rubber squeegee or stiff bristle brush can be used to help remove more of this finish.

After all paint has been removed, flush the entire aircraft off with a pressurized water hose. Let dry. Using clean cotton rags, wipe all surfaces thoroughly with methyl ethyl ketone (MEK).

CORROSION REMOVAL

After paint stripping, any traces of corrosion on the aluminum surface must be removed. *Aircraft Corrosion Control*, a manual published by Aviation Maintenance Publishers, deals with this problem in detail. If corrosion is found, every trace must be removed with fine sandpaper (no emery), aluminum wool, or a Scotch Brite pad. Never use steel wool or a steel brush because bits of steel will become embedded in the aluminum, causing much worse corrosion.

FINISHING PROCEDURE

The following is required for metal pretreatment and painting:

- G-6304 Rand-O-Prep (metal pretreatment)
- 1100 MEK (methyl ethyl ketone)
- G-2404 Randthane epoxy primer (component A)
- G-2405 Randthane epoxy primer (component B)
- G-4201 Randthane primer reducer

For optimum results, temperature and humidity should be within the following limits:

- Relative humidity: 20 percent to 60 percent
- Temperature: Not less than 70°F
- Departure from these limits could result in various application or finish problems

Drying time of the various coatings will vary with temperature, humidity, amount of thinner used, and thickness of paint film.

Safety tips
- Ground the surface you are painting or sanding.
- Do not use an electric drill to mix dope or paint.
- Wear leather-soled shoes in the painting area.
- Wear cotton clothes while painting.
- Keep solvent-soaked rags in fireproof safety container.
- Keep the spray area and floor clean and free of dust buildup.
- Have adequate ventilation. Do not allow mist or fumes to build up in a confined area.
- Do not smoke or have any type of open flame in the area.

Metal pretreatment (aluminum)

In the case of an aircraft that has been stripped of its previous coating, make sure that all traces of paint or paint remover residue have been removed. Give special attention to areas such as seams and around rivet heads. Aircraft should be flushed with plenty of clean water to ensure removal of all contaminants. Let dry. Using clean cotton rags, wipe all surfaces thoroughly with MEK.

Apply G-6304 Rand-O-Prep metal pretreatment liberally with clean rags or a brush to all the aluminum surfaces of the aircraft. While keeping these surfaces thoroughly wet with Rand-O-Prep, scrub briskly with a Scotch Brite pad. It is advisable to wear rubber gloves and to protect your eyes from splashes during this procedure. After the entire aircraft has been treated with this procedure, flush very thoroughly with plenty of clean water. Let dry.

The next step is to thoroughly wipe down the entire aluminum surface with MEK using clean cotton rags. This will ensure all contaminants are removed prior to application of primer.

Primer application

Randthane epoxy primer is mixed as follows: one part by volume of G-2404 Randthane primer (component A); one part by volume of G-2405 Randthane primer (component B); and 1½ parts by volume of G-4201 Randthane primer reducer.

After the three components are thoroughly mixed, let stand for 15–20 minutes prior to starting application. Pot life after mixture is six hours. Randthane primer can be applied using any conventional, electrostatic, or airless type of spray equipment.

Care must be taken that only enough primer be used to prime the surface evenly to about 0.0005 inch (or one-half mil) film thickness. This means that the aluminum substrate should show through with a light yellow coating of the primer coloring the metal.

Drying time of the primer will vary slightly due to differences in temperature and relative humidity at the time of application; but as a general rule, primer should be ready for application of the finish coat within four to six hours.

After the primer is thoroughly dry, wipe entire surface with clean, soft, cotton rags using a little pressure, as if polishing. Next, tack-rag the entire surface. You are now ready for application of the top coat finishing system you have selected.

If more than 24 hours elapse after priming and prior to applying the top coat finish, it is advisable to very lightly scuff the primer with 600 sandpaper and tack-rag before proceeding.

Top coat required materials

The following are required based upon your top coat selection (particular system):

- Randthane polyurethane enamel: selected Randthane color, G-2403 catalyst, and G-4200 reducer
- Randacryl acrylic lacquer: selected Randacrylic color, B-0161 thinner for Randacrylic, and Y-9910 universal Retarder
- Randolph aircraft enamel: selected enamel color and #257 enamel reducer

Top coat finishing systems

Randthane polyurethane enamel top coat application. Thoroughly mix one (1) part by volume of Randthane color to one (1) part by volume G-2403 catalyst. Adding

G-4200 Randthane reducer, thin mixture to 18 to 20 seconds using the #2 Zahn viscosity cup. Let stand for 15 to 20 minutes. Pot life after mixing is approximately six hours, but will vary with color, temperature, and humidity.

Spray a relatively light tack coat on first application. Let dry for at least 15 minutes. Second coat is applied as a full wet cross-coat. Care should be taken that too much paint is not being applied, resulting in runs or sags.

An overnight dry is preferable before taping for trim color application unless forced drying is used. In such a case, 1–2 hours at 140°F is sufficient. After masking and before applying the trim color, lightly scuff the trim color surface using the #400 wet-or-dry sandpaper. Tack-rag and apply trim color. Remove masking tapes as soon as paint has started to set.

Randacrylic acrylic lacquer top coat application. Thoroughly mix Randacryl color with B-0161 Randacrylic thinner as follows: four (4) parts by volume Randacryl color with five (5) parts by volume B-0161 Randacryl thinner. Adjustments to this mixture might be necessary due to spray equipment used or operator technique.

Spray relatively light tack coat on first application. Let dry for approximately 30 minutes. Follow this first coat with at least three full wet cross coats, letting each dry for approximately 30 minutes between coats. If the material is too heavy, orange peel or pinholes are likely to appear.

The gloss in the final coat can be enhanced by adding about a fourth as much Y-9910 Universal Retarder as you have thinner in the material.

An overnight dry is preferable before taping for trim color application. Remove masking tape as soon as paint has started to set.

Randolph aircraft enamel top coat application. Thoroughly mix enamel with only enough #257 enamel reducer to arrive at a viscosity of 25 to 28 seconds using a #2 Zahn viscosity cup.

Spray on a light tack coat. Allow to dry for 15 to 20 minutes, then apply a full, wet cross coat. Allow to dry at least 48 hours before taping for trim colors. After masking and before applying trim color, lightly scuff the trim color surface using #400 wet-or-dry sandpaper. Tack-rag and apply trim color. Remove masking tapes as soon as paint has started to set.

FIBERGLASS REFINISHING

When refinishing any fiberglass component of the aircraft such as radomes, wingtips, antenna, fairings, and the like, it is extremely important that they are protected from paint remover or solvents. The only safe method of removing paint from these components is to sand it off. After paint is removed by sanding, tack-rag the surface and apply a light coat of Randthane primer.

When Randthane primer is dry (four to six hours), briskly wipe the entire surface clean with soft cotton rags (as in polishing). Next, tack-rag the surface and finish in selected finishing system.

SAFER STRIPPING PRODUCT

Most paint strippers have a single chemical in common: metheylene chloride. Additionally, lye or other caustics are commonly added to the strippers, which require special handling to protect skin and the environment. Staz-Wett PB4000 stripper claims to be safe for fiberglass, plastics, metal, skin, and the environment. For further information contact:

Redam Corp.
321 Ross Ave
Hamilton, OH 45013
(800) 234-4583
(513) 863-9393

11

STCs and modifications

AIRCRAFT POPULARITY CAN BE MEASURED by the number of supplemental type certificates issued for a certain make and model, such as Cessna 172 airplanes. When reading the list of STCs, the STC number appears first, followed by the item or part modified by the STC, applicable airplane variants that the STC applies to, and the name and address of the holder of the STC.

AVIONICS AND AUTOPILOTS

SA1-116: Federal F-300 Autopilot (172s); Aircraft Components, Inc., 755 Woodward Ave., Benton Harbor, MI 49023.

SA1-219: Gyro-stabilizer 69A122 (172); Globe Industries, Inc., 125 Sunrise Place, Dayton, OH 45401.

SA1-9 69A115: Gyro-stabilizer; Globe Industries, Inc., 125 Sunrise Pl., Dayton, OH 45401.

SA1-516: Griswold Lateralizer and heading lock; Herbert Haer, 80 Federal Way, Boston, MA 02109.

SA1-612: Installation of Tactair T-101 autopilot and accessory kit; Avionics, Inc., Terminal Bldg, Lunken Airport, Cincinnati, OH 45226.

SA109SW: Automatic pilot AK106 (172); Mitchell Industries, Inc., P.O. Box 610, Municipal Airport, Mineral Wells, TX 76067.

SA1221SW: Mitchell automatic flight system AK312 consisting of Century I with optional omni tracker (172,172 A,B,C,D,E,F,G,H,I,K,L); Century Flight Systems, Inc., F.M. 1195, P.O. Box 610, Mineral Wells, TX 76067.

SA1308WE: Installation of Brittain model CSA-1 Stability Augmentation System; Brittain Industries, P.O. Box 51370 Tulsa, OK 74151.

SA1406SO: Installation of Model CC-1 checkpoint computer (172,172 A,B,C,D,E,F, G,H,I,K,L,M,N,P); Perception Systems, 4500 N Dixie Highway, #C-24, West Palm Beach, FL 33407.

SA1455WE: Installation of Brittain Model CSA-1 Stability Augmentation System; Brittain Industries, P.O. Box 51370, Tulsa, OK 74151.

SA1468WE: Installation of Brittain model B2C Autopilot system; Brittain Industries, P.O. Box 51370, Tulsa, OK 74151.

SA1473WE: Installation of Brittain Model B2C Flight Control System; Brittain Industries, P.O. Box 51370, Tulsa, OK 74151.

SA147CE: Moni-Meters in aircraft equipped with King KY-90, KY-85, Narco Mark 4 or Mark 12 transceivers; Mobile Engineering, 220 Southdale Center, Minneapolis, MN 55435.

SA1568CED: Installation of KAP 100 single-axis KAP 150 two-axis or KFC 150 two-axis flight control system (172 RG); King Radio Corp., 400 N. Rogers Rd., Olathe, KS 66062.

SA1570CED: Installation of KAP 100 single-axis, KAP 150 two-axis, or KFC 150 two-axis flight control system (172 P); King Radio Corporation, 400 North Rogers Rd., Olathe, KS 66062.

SA1645SW: Mitchell automatic flight system AK403, consisting of Century IIB autopilot with optional radio coupler (172 D through 172 I,K,L); Century Flight Systems, Inc., F.M. 1195, P.O. Box 610, Mineral Wells, TX 76067.

SA168GL: Install Heath Kit Model 01-1154 aircraft digital clock/timer (172 N); The Heath Company, Hilltop Road, Benton Harbor, MI 49022.

SA168GL: Installation of Heath Kit Model 01-1154 Aircraft Digital Clock/Timer (172 N); The Heath Company, Hilltop Rd., Benton Harbor, MI 49022.

SA1798SW: Mitchell Automatic Flight System AK467 consisting of Century IIB Autopilot with optional Radio Coupler (172 M); Century Flight Systems, Inc., F.M. 1195, P.O. Box 610, Mineral Wells, TX 76067.

SA1799SW: Mitchell Automatic Flight System AK472 consisting of Century I Autopilot with optional omni tracker (172 M); Century Flight Systems, Inc., F.M. 1195, P.O. Box 610, Mineral Wells, TX 76067.

SA17EA: Installation of Tactair T-101 autopilot and optional installation of Tactair T-201 autopilot accessory kit (for use with Tactair T-101 autopilot) (172 D,P172 D); Avionics, Inc., Terminal Building, Lunken Airport, Cincinnati, OH 45226.

SA1806WE: Installation of Brittain Industries B2C Flight Control system; Brittain Industries, P.O. Box 51370, Tulsa, OK 74151.

SA232EA: Installation of Tactair omni-lock model OL-1 and optional installation of Tactair localizer adapter Model LA-1 (for use with Tactair model omni-lock model OH-4) (172 B,C); Avionics, Inc., Terminal Building, Lunken Airport, Cincinnati, OH 45226.

SA299GL: Installation of Cessna Nav-O-Matic 200A autopilot (172 G); Galesburg AP and L, R.R. 2, Airport, Galesburg, IL 61401.

SA2619WE: Installation of Pathfinder model P1 autopilot (172 D through M); Astronautics Corporation of America, 2416 Amsler Street, Torrance, CA 90505.

SA2620WE: Installation of Pathfinder Model P2 autopilot system (172 C through L); Astronautics Corp. of America, 2416 Amsler Street, Torrance, CA 90505.

SA2654WE: Installation of Pathfinder model P2A autopilot (172 D,E,F,G,H,I,K,L,M); Astronautics Corp. of America, 2416 Amsler Street, Torrance, CA 90505.

SA3-90: Lear ADF Loop and housing; Capitol Aviation, Inc., Capitol Airport, Springfield, IL 62705.

SA3-108: Single-axis automatic pilot A-2 or A-3; Javelin Aircraft Co., Inc., 9175 East Douglas, Wichita, KS 67207.

SA3046SWD: Mitchell automatic flight system AK524 consisting of Century IIB autopilot with optional radio coupler (172 D,E,F,G,H,I,K,L); Century Flight Systems, Inc., FM 1195, P.O. Box 610, Mineral Wells, TX 76067.

SA313SO: Federal FS103 autopilot; A and E Service, McCollum Airport, Marietta, GA 30060.

SA3200SWD: Automatic flight system AK467 consisting of Century II autopilot with optional radio coupler (172 M,N); Century Flight Systems, Inc., F.M. 1195, P.O. Box 610, Mineral Wells, TX 76067.

SA3201SWD: Automatic flight system AK472 consisting of Century I autopilot with optional omni tracker (172 M,N); Century Flight Systems, Inc., F.M. 1195, P.O. Box 610, Mineral Wells, TX 76067.

SA3201SWD: Automatic flight system AK472 consisting of Century I autopilot with optional omni tracker (172 M,N); Century Flight Systems, Inc., F.M. 1195, P.O. Box 610, Mineral Wells, TX 76067.

SA3209NM: Installation of SBI Model CFS-1000A,1001A, FT-100, or FT-101 fuel flow indicating system (172 I,K,L,M); Symbolic Displays, Inc., P.O. Box 19626, Irvine, CA 92713.

SA3227SWD: Mitchell automatic flight system AK699, consisting of Century I autopilot with optional omni tracking system (172 RG); Century Flight Systems, Inc., F.M. 1195, P.O. Box 610, Mineral Wells, TX 76067.

SA3228SWD: Mitchell automatic flight system AK698, consisting of Century IIB autopilot with optional radio coupler (P172 D); Century Flight Systems, Inc., F.M. 1195, P.O. Box 610, Mineral Wells, TX 76067.

SA3284SWD: Installation of Mitchell automatic flight system AK730, consisting of Century I autopilot with optional omni tracking system (R172 K); Century Flight Systems, Inc., F.M. 1195, P.O. Box 610, Mineral Wells, TX 76067.

SA3285SWD: Installation of Mitchell automatic flight system AK731, consisting of Century IIB autopilot with radio coupler (R172 K); Century Flight Systems, Inc., F.M. 1195, P.O. Box 610, Mineral Wells, TX 76067.

SA3286SWD: Installation of Mitchell automatic flight system AK732, consisting of Century III autopilot with optional radio and glide slope coupler (R172 K); Century Flight Systems, Inc., F.M. 1195, P.O. Box 610, Mineral Wells, TX 76067.

SA3301SWD: Mitchell automatic flight system AK730, consisting of Century I autopilot with optional omni tracking system (172 N); Century Flight Systems, Inc., F.M. 1195, P.O. Box 610, Mineral Wells, TX 76067.

SA3302SWD: Mitchell automatic flight system AK731 consisting of Century IIB autopilot with optional radio coupler (172 N); Century Flight Systems, Inc., F.M. 1195, P.O. Box 610, Mineral Wells, TX 76067.

SA3303SWD: Mitchell automatic flight system AK732 consisting of Century III autopilot with optional radio and glide slope couplers (172 N); Century Flight Systems, Inc., F.M. 1195, P.O. Box 610, Mineral Wells, TX 76067.

SA3363SW: Mitchell automatic flight system AK861 consisting of Century 21 Autopilot (172 RG); Century Flight Systems, Inc., F.M. 1195, P.O. Box 610, Mineral Wells, TX 76067.

SA3417SWD: EDO Avionics automatic flight system AK904 consisting of Century 31 autopilot (172 P); Century Flight Systems, Inc., F.M. 1195, P.O. Box 610, Mineral Wells, TX 76067.

SA3418SWD: EDO Avionics automatic flight system AK920 consisting of Century 21 autopilot (172 P); Century Flight Systems, Inc., F.M. 1195, P.O. Box 610, Mineral Wells, TX 76067.

SA3425SWD: Installation of Century automatic flight system Model AK928 consisting of Century 31 autopilot (172 D-N); Century Flight Systems, Inc., F.M. 1195, P.O. Box 610, Mineral Wells, TX 76067.

SA3426SWD: Installation of Century automatic flight system Model AK929 consisting of Century 21 autopilot (172 D-N); Century Flight Systems, Inc., F.M. 1195, P.O. Box 610, Mineral Wells, TX 76067.

SA3470SWD: Installation of Century automatic flight system Model AK977 consisting of Century 2010/2030/2031 autopilot (172 D-N); Century Flight Systems, Inc., F.M. 1195, P.O. Box 610, Mineral Wells, TX 76067.

SA3486SWD: Installation of Century automatic flight system Model AK993 consisting of Century 2110-2130/2131 autopilot (172 P); Century Flight Systems, Inc., F.M. 1195, P.O. Box 610, Mineral Wells, TX 76067.

SA380GL: Installation of Tull microwave landing system airborne equipment (172 M,N); Burlington Northern Automotive, 3600 East 70th Street, Minneapolis, MN 55450.

SA3824SWD: Mitchell automatic flight system AK730 consisting of Century I autopilot with optional omni tracking system (172 K); Mitchell Industries, Inc., P.O. Box 610, Mineral Wells, TX 76067.

SA393WE: Pneumatic pitch assist PC-1 and optional altitude hold AH-1 (172); Brittain Industries, Inc., P.O. Box 51370, Tulsa, OK 74151.

SA4-531: Federal F-200 autopilot; Aircraft Components, Inc., 755 Woodward Ave., Benton Harbor, MI 49023.

SA4-40: Automatic rudder control; Motorola Aviation Electronics, Inc., 3302 Airport Ave., Santa Monica, CA 90406.

SA4-602: Autopilot; Brittain Industries, Inc., 12027 South Prairie Ave., Hawthorne, CA 90250.

SA4-367: Super automatic rudder control; Motorola Aviation Electronics, Inc., 3302 Airport Ave., Santa Monica, CA 90406.

SA4-676: RDF-2 radio direction finder; Savage Industries, Inc., 3125 N. 29 Ave., Phoenix, AZ 85000.

SA4-925 Magnetic flight director 2000; Clarkson Co., Paul Spur, AZ.

SA411SW: Automatic pitch trim AK138 (172); Mitchell Industries, Inc., P.O. 610, Municipal Airport, Mineral Wells, TX 76067.

SA4197SW: S-TEC System 60 (two-axis) flight guidance system, Model ST-011, with optional flight director/steering horizon and vertical speed indicator/selector (14 volt system) (172 P,Q); S-TEC Corporation, Route 3, Building 946, Wolters Industrial Complex, Mineral Wells, TX 76067.

SA4221SW: S-TEC System 60 (two-axis) flight guidance system, Model ST-019, with optional flight director steering horizon and vertical speed selector (14 volt system) (172 M,N); S-TEC Corporation, Route 3, Building 946, Wolters Industrial Complex, Mineral Wells, TX 76067.

SA4230SW: S-TEC 60-roll flight guidance kit, Model ST-008 (172 M,N); S-TEC Corporation, Route 3, Building 946, Wolters Industrial Complex, Mineral Wells, TX 76067.

SA4243SW: S-TEC System 60 (two-axis) flight guidance system Model ST-025 with optional flight director/steering horizon and optional vertical speed/indicator selector (28 volt system) (172 N); S-TEC Corporation, Route 3, Building 946, Wolters Industrial Complex, Mineral Wells, TX 76067.

SA4244SW: S-TEC System 60 single-axis flight guidance system, Model ST-012 (28 volt system) (172 N); S-TEC Corporation, Route 3, Building 946, Wolters Industrial Complex, Mineral Wells, TX 76067.

SA4338SW: S-TEC pitch stabilization system, Model ST-042 (14 volt system) (172 M,N); S-TEC Corporation, Route 3, Building 946, Wolters Industrial Complex, Mineral Wells, TX 76067.

SA4417SW: Installation of S-TEC System 60 stabilization system ST-060 (28 volt) (172 N); S-TEC Corporation, Route 3, Building 946, Wolters Industrial Complex, Mineral Wells, TX 76067.

SA4424SW: Installation of S-TEC System 60, two-axis flight guidance system Model FT-046 with optional auto-trim system, optional flight director/steering horizon (172 RG); S-TEC Corporation, Route 3, Building 946, Wolters Industrial Complex, Mineral Wells, TX 76067.

SA4458SW: Installation of S-TEC System 60, single-axis flight guidance system Model ST-045 (172 RG); S-TEC Corporation, Route 3, Building 946, Wolters Industrial Complex, Mineral Wells, TX 76067.

SA4477SW: Installation of S-TEC System 60-pitch stabilization system, Model ST-062 with optional auto-trim system (172 RG); S-TEC Corporation, Route 3, Building 946, Wolters Industrial Complex, Mineral Wells, TX 76067.

SA4536NM: Installation of modified instrument panel and control yoke; DelAir, 2121 S. Wildcat Way, Porterville, CA 93257.

SA4717SW: Installation of Texas Instruments Model 9100 LORAN C navigator (172, 172 A,B,C,D,E,F,G,H,I,L,M,N,P); Texas Instruments, P.O. Box 405 M/S 3439, Lewisville, TX 75067.

SA494SO: Installation of radio equipment in instrument panel (172 L); Burnside-Ott Aviation Training Center, 12800 S.W. 137th Avenue, Miami, FL 33156.

SA5111SWD: S-TEC System 60 two-axis automatic flight guidance system ST-095 (172 D,E,F,G,H,I,K,L); S-TEC Corporation, Route 4, Building 946, Wolters Industrial Complex, Mineral Wells, TX 76067.

SA5113SWD: S-TEC System 60-pitch stabilization system ST-096 (172 D,E,F,G,H,I,K,L); S-TEC Corporation, Route 4, Building 946, Wolters Industrial Complex, Mineral Wells, TX 76067.

SA5124SWD: S-TEC System 60 single-axis autopilot (14 volt) (172, 172 A,B,C,D,E,F,G,H,I,K,L); S-Tec Corporation, Route 4, Building 946, Wolters Industrial Complex, Mineral Wells, TX 76067.

SA5133SWD: S-TEC System 60 two-axis automatic flight guidance system ST-019 with optional flight director steering horizon (172 M,N); S-TEC Corporation, Route 3, Building 946, Wolters Industrial Complex, Mineral Wells, TX 76067.

SA5139SWD: S-TEC System 60 two-axis automatic flight guidance system ST-025 with optional flight director steering horizon (172 N,P); S-TEC Corporation, Route 3, Building 946, Wolters Industrial Complex, Mineral Wells, TX 76067.

SA5150SWD: S-TEC System 60-pitch stabilization ST-042 (172 M,N); S-TEC Corporation, Route 3, Building 946, Wolters Industrial Complex, Mineral Wells, TX 76067.

SA5151SWD: S-TEC System 60 pitch-stabilization system ST-060 (172 N,P); S-TEC Corporation, Route 3, Building 946, Wolters Industrial Complex, Mineral Wells, TX 76067.

SA5188SWD: S-TEC System 60 pitch-stabilization system ST-062 with optional automatic electric system (172 RG); S-TEC Corporation, Route 3, Building 946, Wolters Industrial Complex, Mineral Wells, TX 76067.

SA5192SWD: S-TEC 40/50 single and two-axis automatic flight guidance systems ST-184-40/50 (172 M,N,P); S-TEC Corporation, Route 3, Building 946, Wolters Industrial Complex, Mineral Wells, TX 76067.

SA5195SWD: S-TEC System 40/50 single and two-axis automatic flight guidance systems, ST-183-40/50 (172 M,N); S-TEC Corporation, Route 3, Building 946, Wolters Industrial Complex, Mineral Wells, TX 76067.

SA5200SWD: S-TEC System 40 single-axis automatic flight guidance system ST-182-40 (172,172 A,B,C,D,E,F,G,H,I,K,L); S-TEC Corporation, Route 4, Building 946, Wolters Industrial Complex, Mineral Wells, TX 76067.

SA5201SWD: S-TEC System 50 two-axis automatic flight guidance system, ST-182-50 (172 D,E,F,G,H,I,K,L); S-TEC Corporation, Route 4, Building 946, Wolters Industrial Complex, Mineral Wells, TX 76067.

SA5218SWD: S-TEC System 40/50 single and two-axis automatic flight guidance system, ST-207-40/50 (172 RG); S-TEC Corporation, Route 4, Building 946, Wolters Industrial Complex, Mineral Wells, TX 76067.

SA5235SWD: S-TEC manual electric trim system (172 RG); S-TEC Corporation, Route 4, Building 946, Wolters Industrial Complex, Mineral Wells, TX 76067.

SA5247SWD: S-TEC System 60 two-axis automatic flight guidance system ST-025 with optional automatic electric trim (R172 K); S-TEC Corporation, Route 4, Building 946, Wolters Industrial Complex, Mineral Wells, TX 76067.

SA5284SWD: Installation of S-TEC System 40/50 single and two-axis automatic flight guidance system ST-184-40/50 (R172 K); S-TEC Corporation, Route 4, Building 946, Wolters Industrial Complex, Mineral Wells, TX 76067.

SA5287SWD: Installation of S-TEC System 60 two-axis automatic flight guidance system ST-019 with optional automatic electric trim (R172 J,K); S-TEC Corporation, Route 3, Building 946, Wolters Industrial Complex, Mineral Wells, TX 76067.

SA5291SWD: Installation of S-TEC System 60 single-axis automatic flight guidance system ST-012 (172 M,N,P,Q); S-TEC Corporation, Route 3, Building 946, Wolters Industrial Complex, Mineral Wells, TX 76067.

SA5292SWD: Installation of S-TEC System 60 single-axis automatic flight guidance system ST-012 (R172 K); S-TEC Corporation, Route 3, Building 946, Wolters Industrial Complex, Mineral Wells, TX 76067.

SA5297SWD: Installation of S-TEC System 60 pitch stabilization system ST-060 (R172 K); S-TEC Corporation, Route 3, Building 946, Wolters Industrial Complex, Mineral Wells, TX 76067.

SA5309SWD: Installation of S-TEC System 60 single-axis automatic flight guidance system Model ST-008 (172 M,N); S-TEC Corporation, Route 3, Building 946, Wolters Industrial Complex, Mineral Wells, TX 76067.

SA5309SWD: Installation of S-TEC System 60 single-axis automatic flight guidance system Model ST-008 (172 M,N); S-TEC Corporation, Route 3, Building 946, Wolters Industrial Complex, Mineral Wells, TX 76067.

SA5310SWD: Installation of S-TEC System 60 single-axis automatic flight guidance system Model ST-008 (R172 J,K); S-TEC Corporation, Route 3, Building 946, Wolters Industrial Complex, Mineral Wells, TX 76067.

SA5314SWD: Installation of S-TEC System 40/50 single and two-axis automatic flight guidance system Model ST-183-40/50 (R172 J,K); S-TEC Corporation, Route 3, Building 946, Wolters Industrial Complex, Mineral Wells, TX 76067.

SA5322SWD: Installation of S-TEC System 60-pitch stabilization system Model ST-042, with optional automatic electric trim system (R172 J,K); S-TEC Corporation, Route 3, Building 946, Wolters Industrial Complex, Mineral Wells, TX 76067.

SA5371SWD: Installation of S-TEC manual electric trim system Model ST-157 (172 M-Q); S-TEC Corporation, Route 3, Building 946, Wolters Industrial Complex, Mineral Wells, TX 76067.

SA5373SWD: Installation of S-TEC manual electric trim system Model ST-157 (R172 K); S-TEC Corporation, Route 3, Building 946, Wolters Industrial Complex, Mineral Wells, TX 76067.

SA5377SWD: Installation of S-TEC System 60 two-axis automatic flight guidance system, Model ST-075, with optional automatic electric trim system (R172 E-H); S-TEC Corporation, Route 3, Building 946, Wolters Industrial Complex, Mineral Wells, TX 76067.

SA5379SWD: Installation of S-TEC System 60-pitch stabilization system, Model ST-096, with optional automatic electric trim system (R172 E-H); S-TEC Corporation, Route 3, Building 946, Wolters Industrial Complex, Mineral Wells, TX 76067.

SA5381SWD: Installation of S-TEC System 60 single-axis automatic flight guidance system, Model ST-094 (R172 E-H); S-TEC Corporation, Route 3, Building 946, Wolters Industrial Complex, Mineral Wells, TX 76067.

SA5383SWD: Installation of S-TEC System 40/50 single and two-axis automatic flight guidance systems, Model ST-182-40/50 (R172 E-H); S-TEC Corporation, Route 3, Building 946, Wolters Industrial Complex, Mineral Wells, TX 76067.

SA603SW: Installation of Mitchell automatic flight system model AK-191 consisting of Century II and radio coupler and model AK192 automatic aileron stabilizer (172 D,E,F,G,H,I,K,L); Century Flight System, Inc., F.M. 1195, P.O. Box 610, Mineral Wells, TX 76067.

SA7156SWD: Installation of S-TEC System 60 two-axis automatic flight guidance system, Model ST-046 with optional automatic electric trim system (172 RG); S-TEC Corp., Rte 4, Building 946, Wolters Complex, Mineral Wells, TX 76067.

SA7188SWD: Installation of S-TEC manual electric trim system Model ST-447 (172 M,N); S-TEC Corporation, Route 3, Building 946, Wolters Industrial Complex, Mineral Wells, TX 76067.

SA7189SWD: Installation of S-TEC manual electric trim system Model ST-447 (R172 J,K); S-TEC Corporation, Route 3, Building 946, Wolters Industrial Complex, Mineral Wells, TX 76067.

SA808SW: Installation of Mitchell omni tracker Model AK246; Century Flt. Systems, Inc., 227 Oregon St., El Segundo, CA 90245.

SA974EA: Installation of L/R wingtip transparent fairing and sensor mounting bracket (172 M); Rock Avionics Systems, Inc., 412 Avenue of the Americas, New York, NY 10011.

ENGINE MODIFICATIONS

SA1-124: Lube oil filter BP55-1; Fram Corp., 105 Pawtucket Ave., Providence, RI 02916.
SA1087GL: Installation of a Shadin Company fuel flow indicating system (172 RG); Shadin Co., Inc., 14280 N. 23rd Ave., Plymouth, MN 55447.

SA10WE: Arens Vernier type throttle control (172); Van Nuys Skyways, Inc., 16700 Roscoe Blvd, Van Nuys, CA 91408.

SA116EA: Installation of Fram Part Flow Lube Oil Filter Model PB 55-1 (172); Fram Aerospace Division of Fram Corporation, 750 School Street, Pawtucket, RI 02860.

SA117EA: Flow lube oil filter PB 55-1; Fram Aerospace, Division of Fram Corp., 750 School St., Pawtucket, RI 02860.

SA1315WE: Installation of Tri-Star Corp. mixture monitor (exhaust gas temperature probe, gauge, and selector); Universal Corp., 730 Independent Ave., Grand Junction, CO 81505.

SA135EA: Installation of engine exhaust combustion monitor (172, 172 A,B,C,D); Rosemount Engineering Co., 12001 West 78th St., Eden Prairie, MN 55343.

SA1418SO: Replacement of rotor vanes in Airborne 211CC and 212CW vacuum pumps; U.S. Air Source, Ltd., 3640 Atlanta Hwy., Athens, GA 30604.

SA1512WE: Installation of exhaust gas temperature monitoring system Model EGT-1; K. S. Avionics, 18145 Judy Street, Castro Valley, CA 94546.

SA157NE: Installation of Graphic Engine Monitor System model GEM- GEM-602 S/N 403 and subsequent; Insight Instrument Corp., Box 194, Ellicott Station, Buffalo, NY 14205.

SA1626NM: Installation of Electronic Int'l EGT/CHT instrument and accessories; Electronic's Int'l., Inc., 5289 NE Elam Young Pkwy G200, Hillsboro, OR 97124.

SA2350NM: Installation of Electronic Int'l digital carburetor/outside air temperature gauges; Electronic's Int'l., Inc., 5289 NE Elam Young Pkwy G200, Hillsboro, OR 97124.

SA1686WE: Installation of engine mounts P/N VIP50525-1 and VIP50545-1 (P172 D); Vibration Isolation Products Corp., 11275 San Fernando Road, San Fernando, CA 91340.

SA18EA: Wet-to-dry vacuum pump conversion kit 300-1 (172 A,B,C,D,E,F,G,H); Airborne Manufacturing Company, 711 Taylor Street, Elyria, OH 44035.

SA1947CE: Replace existing engine driven vacuum pump with engine driven piston-type vacuum pump (172 through 172 Q); Sigma-Tek, Inc., 1326 South Walnut Street, Wichita, KS 67213.

SA2105WE: Installation of Model EGT-3 exhaust gas temperature monitor (with rising temperature alarm); K. S. Avionics, 18145 Judy Street, Castro Valley, CA 94546.

SA2190NM: Installation of Electronic Int'l Digital Volt/AMP Gauges and accessories; Electronic's Int'l., Inc., 5289 NE Elam Young Pkwy G200, Hillsboro, OR 97124.

SA2557NM: Installation of KS Avionics EGT/CHT-2 combines exhaust gas and cylinder head temperature monitoring system; KS Avionics, Inc., 25216 Cypress Ave., Hayward, CA 84544.

SA2586NM: Installation of J.P. Instruments exhaust gas and/or cylinder head temperature monitoring system; J.P. Instruments, P.O. Box 7033, Huntington Beach, CA 92615.

SA2687NM: Installation of Turboplus/Electronics Int'l digital temperature gauges and switches to reflect new Turboplus face plates to redesignate selected functions or switch positions; Turboplus, Inc., Tacoma Narrows Airport, 1520 26th Ave., NW, Gig Harbor, WA 98335.

SA2693NM: Installation of Electronic Int'l digital volt/amp gauges and accessories; Electronic's Int'l., Inc., 5289 NE Elam Young Pkwy G200, Hillsboro, OR 97124.

SA271NW: Removal of paper air induction filter (172); Kenmore Air Harbor, P.O. Box 64, Kenmore, WA 98028.

SA3-664: Install model 113A6, 200CW-6 or 212CW-6 dry vacuum pump and instrument vacuum system (172, 172 A,B,C,D,E,F); Parker Hannifin Corp., Airborne Division, 711 Taylor St., Elyria, OH 44035.

SA308EA: Installation of A.R.P. Industries, Inc., carburetor ice detection system 105AP in airplanes equipped with engines using certain Marvel-Schebler carburetors; Alfred R. Puccinelli, DER 1-145, 36 Bay Drive East, Huntington, LI NY 11743.

SA316SO: Carburetor ice detector; C.B. Shivers Jr., 8928 Valleybrook Rd., Birmingham, AL 35206.

SA3659SW: Installation of Jetton Alert II alarm system; Jetton Aircraft, P.O. Box 187, Addison, TX 75001.

SA3727SA: Installation of vacuum pump cooling shroud (172 I,K,L,M,N,P); RAM Aircraft Corp., P.O. Box 5219, Waco, TX 76708.

SA3777WE: Installation of flexible stainless steel oil cooler hoses (172 I,K through N); Aircraft Metal Products Corporation, 4206 Glencoe Avenue, Venice, CA 90291.

SA3796WE: Installation of Elano P/N 099001-063 F, or later FAA-approved revision muffler in lieu of original Avcon muffler (172, 172 A through 172 H (landplane only, normal category); Del-Air, P.O. Box 746, Strathmore, CA 93267.

SA3833: Installation of Oilamatic Engine Preoiler (172 I,K,L,M,N,P); Oilamatic, Inc., P.O. Box 5284, Englewood, CO 80155.

SA3858WE: Installation of an air/oil separator (172 I,K,L,M,N); Walker Engineering Company, 5760 West 3rd Street, Los Angeles, CA 90036.

SA3862NM: Installation of Electronic Int'l digital TIT/EGT CHT instruments and accessories; Electronic's Int'l., Inc., 5289 NE Elam Young Pkwy G200, Hillsboro, OR 97124.

SA398NE: Installation of Auto-Vac-2 manifold vacuum system; Raymond W. Ives, RFD 1, Box 141A, Pomfret Center, CT 06259.

SA4-686: Full flow lube oil filter 30409A with element 1A0235; Winslow Aerofilter Corp., 4069 Hollis St., Oakland, CA 94608.

SA4166NM: Installation of Fairchild voltage regulator No. UA78 GUIC to replace the original voltage regulator in Cessna turn coordinator P/N 0661003-0506; Paul Malkasian, 1036 Euclid Ave., Edmonds, WA 98020.

SA4172NM: Installation of Oberg 600 series oil filter (172 I-Q); Aviation Development Co., 1304 NW 200th, Seattle, WA 98177.

SA429WE: Exhaust pipe extensions; Lloyd M. Green, 619 W. Roses Road, San Gabriel, CA 91776.

SA4302NM: Installation of Electronic Int'l Digital Model SR-8 or Model US-8 Digital Automatic Engine Analyzer; Electronic's Int'l., Inc., 5289 NE Elam Young Pkwy G200, Hillsboro, OR 97124.

SA489EA: Installation of A.R.P. Industries, Inc., carburetor ice detection system 105AP in airplanes equipped with engines using certain Marvel-Schebler carburetors; Alfred R. Puccinelli, DER 1-145, 36 Bay Drive East, Huntington, LI NY 11743.

SA503NE: Installation of oil cooler door (172 K-M); E & M Aero Products, Inc., 20 Woodslee Ave., Paris, Ont. N3L3N6 Canada.

SA508NE: Installation of Sperry Model MLZ-900 microwave landing system (172 M); MSI 600 Maryland Ave., Suite 695, Washington, DC 20024.

SA5621SW: Installation of electrically driven vacuum pump (172 M,N,P,Q); Aero Safe Corporation, P.O. Box 10206, Fort Worth, TX 76114.

SA5887SW: Installation of electrically driven vacuum pump as a standby auxiliary pump to the existing instrument air system (172 RG); Aero Safe Corp., P.O. Box 10206, Fort Worth, TX 76114.

SA589WE: Installation of full-flow lube oil filters on aircraft for the benefit of the engine; Worldwide Aircraft Filter Corp., 1685 Abram Ct., P.O. Box 175B, San Leandro, CA 94577.

SA596EA: Installation of electronic voltage alarm drawing No. 80104; McKinley Engineering Corp., P.O. Box 275, Palisades Park, NJ 07650.

SA630WE: Installation of oil filter assemblies; Nelson, Div. of Nelson Ind. Inc., P.O. Box 280, Stoughton, WI 53589.

SA637NW: Installation of AC oil filter assembly OF-104A; Turbotech, Inc., P.O. Box 61586, Vancouver, WA 98666.

SA695CE: Install Frantz oil filter; Schmidt Aero Service, Municipal Airport, Worthington, MN 56187.

SA702GL: Installation of a cooling shroud on engine-driven dry air pumps (172 K,L,M,N,P,Q); S & M Products, 2515 East Bonnie Brook Lane, Waukegan, IL 60087.

SA71GL: Replace existing engine air filter frame assembly with Brackett Aircraft frame assembly; Brackett Aircraft Specialties, 9600 West 52nd St, Kenosha, WI 53140.

SA751EA: Installation of winterization kit (172 B,G); Franklin Engine Company, Inc., Old Liverpool Road, Syracuse, NY 13206.

SA798EA: Installation of Monitair exhaust gas combustion monitor kit; Rosemount Engineering Co., 12001 West 78th St., Eden Prairie, MN 55343.

SA86NW: Installation of automatic carburetor alternate air flow control unit; Aero-Deicers, Inc., 5407 S.E. 62nd Ave., Portland, OR 97206.

SA965CE: Installation of Model IU328-001 vacuum pump and IU292-003 suction relief valve (172 L serial numbers 17259224 through 17260758); Edo-Aire Wichita Division, 1326 South Walnut, Wichita, KS 67213.

PERFORMANCE MODIFICATIONS

SA1-438: Lycoming O-360-A1A engine and McCauley propeller with Woodward governor; McCauley Industrial Corp., 1840 Howell Ave., Dayton, OH 45401.

SA1078WE: Franklin 6A-335-B engine with McCauley 2A31C/84S-6 propeller using EDO 89-2000 floats; Columbia Marine, Inc., P.O. Box 179, Vancouver, WA 98663.

SA121CE: Nose and main gear wheel fairing (Jetstreams) using standard unaltered Cessna landing gear and nose gear fork for model and year using 500 x 5 nose and 600 x 6 main gear tires; Creative Designs, 1338 Orkla Dr., Minneapolis, MN 55427.

SA1221SO: Fabrication and installation of wheel fairings; Windy's Aircraft Parts, Div. of Southern Avn. of Laurel, Inc., P.O. Box 6408, Laurel, MI 39441.

SA1225CE: Installation of Lycoming O-320-E2D/STC SE1226CE or O-320-D2G engine and McCauley 1C160/CTM 7557 or 1C160/DTM 7557 propeller (172 I,K,L,M landplane); Schneck Aviation, Inc., Greater Rockford Airport, P.O. Box 6417, Rockford, IL 61125.

SA1294SO: Installation of an adjustable nine-inch trim tab on left aileron (most 172); Aero-Trim, Inc., 1130 102nd St., Bay Harbor, FL 33407.

SA1295SO: Installation of an adjustable nine inch trim tab on left aileron (172 RG); Aero-Trim, Inc., 1130 102nd St., Bay Harbor, FL 33154.

SA12WE: Fiberglass wingtips; Met-Co-Aire, P.O. Box 2216, Fullerton, CA 92633.

SA1324CE: Installation of Lycoming O-360-A1A engine and Hartzell HC-C2YK-1B/7666A-2 propeller (172 N); Avcon Industries, Inc., 1006 West 53rd Street North, P.O. Box 4248, North Wichita Station, Wichita, KS 67204.

SA1334WE: Installation of Franklin 6A-350-C2 engine and McCauley 2A31C/84S-6 propeller (172 B through G); Columbia Marine, Inc., P.O. Box 179, Vancouver, WA 98663.

SA1342GL: Installation of Porsche Model PFM3200/NO3 engine and Hartzell propeller (172 N); Porsche Aviation Products, Inc., Galesburg Airport, RR 2, Box 118A, Galesburg, IL 61401.

SA1356GL: Installation of Lycoming O-320-B2J,D2G, or D1A engine (172 N); Penn Aero Service Inc., 2499 Bath Rd., Penn Yan, NY 14527.

SA1371SW: Wing leading cuff, wingtips and upper surface flow fences; Barbara or Bob Williams, Box 431, 213 North Clark, Udall, KS 67146.

SA1402CE: Install wing stall fences and aileron gap seals (172 M,N); Barbara or Bob Williams, Box 431, 213 North Clark, Udall, KS 67146.

SA1431WE: Installation of rudder trim system; Robertson Aircraft Corp., 839 W. Perimeter Rd., Renton, WA 98055.

SA1437CE: Continental IO-360-K engine installation rerated and designated IO-360-KC/STC SE1436CE (R172 K); Brad B. Isham, 416 W. 4th St., Valley Center, KS 67147.

SA1473SO: Installation of Lycoming O-320-D2A engine (172 N); Max Acft. Engine Service, Inc., Wetumpka Muni. Apt, Wetumpka, AL 36092.

SA1474SW: Installation of MASA 1223000-7 and -8 wingtip; Barbara or Bob Williams, Box 431, 213 North Clark, Udall, KS 67146.

SA1484CE: Installation of flap and aileron seals (172, 172 A,B,C,D,E,F,G,H,I,K,L,M, N,P); Aircraft Development Company, 1326 North Westlink Blvd, Wichita, KS 67212.

SA1491CE: Installation of flap and aileron gap seals (R172 K and 172 RG); Aircraft Development Co., 1326 N. Westlink Blvd., Wichita, KS 67212.

SA150NW: Installation of flap gap seals; Robert L. or Barbara V. Williams, Box 431, Udall, KS 67146.

SA158CE: Cessna wingtips and attached lights or equivalent; Schott Aviation, Inc., Baldwin Field, Quincy, IL 62301.

SA1665CE: Installation of strut/wing and strut fuselage fairings, lightening hole covers, and aerodynamic putty (172, 172 A through 172 G landplanes and floatplanes, 172, 172 H through 172 I landplanes only); Aircraft Development Company, 1326 North Westlink Boulevard, Wichita, KS 67212.

SA1677SW: Wing leading edge cuff; Southern Aircraft Corp., P.O. Box 78, Page, OK 74939.

SA1689WE: Installation of drooped ailerons in flaps down mode, contoured wing leading edge and wing stall fences; Sierra Industries, Inc., P.O. Box 5184, Uvalde, TX 78802.

SA172NW: Installation of drooped type wingtips; Robert C. Cransdale, 29511 9th Place S., Federal Way, WA 98002.

SA175EA: Installation of wing struts modified to include Springs Model 200 hydraulic-pneumatic shock absorbers; Wings With Springs, Inc., Box 495, R.D. 3, Latrobe, PA 15650.

SA1769: Installation of Lycoming O-320-D2J engine (172 N); Penn Yan Aero Service, Inc., Product Dev. Div., RR 1, Box 165, Mahomet, IL 61853.

SA1774WE: Installation of Lycoming O-320-E2D engine and McCauley 1C172MTM7653 propeller (172 B through H); Columbia Marine, Inc., P.O. Box 179, Vancouver, WA 98663.

SA181GL: Installation of McCauley 1A200/DFA propeller (172 A seaplane); Les G. Taylor, 1115 19th Street South, St. Cloud, MN 56301.

SA1895NM: Installation of flap gap fairing (172 RG); Thermal Aircraft Co., P.O. Box 133, Aguila, AZ 85320.

SA1945CE: Increase normal category maximum gross weight from 2300 to 2400 lbs and increase utility category maximum gross weight from 2000 to 2100 lbs (172 N); Ralph W. Rissmiller Jr., 9726 W. 9th St. Airport, Wichita, KS 67212.

SA2002WE: Installation of recontoured wing leading edge, stall fences, wingtips and positive aileron seals (172 through 172 L); Sierra Industries, Inc., P.O. Box 5184, Uvalde, TX 78802.

SA204GL: Installation of AVCO Lycoming O-320-E2D engine (172 H); Professional Aviation, Inc., Porter County Airport, 3801 Murvihill Rd., Valparaiso, IN 46383.

SA2075: Installation of Madras Air Service Inc. Super Tips (172, 172 A through 172 L normal category); Madras Air Service, Inc., Route 2, Box 1225, Madras, OR 97741.

SA2137SO: Fabrication and installation of nosewheel fairings; Windy's Aircraft Parts, Div. of Southern Avn. of Laurel, Inc., P.O. Box 6408, Laurel, MI 39441.

SA2196CE: Installation instructions for gross weight increase (172 K-P); Air Plains Svcs. Corp., P.O. Box 541, Wellington Airport, Wellington, KS 67152.

SA2268WE: Installation of Brittain Industries B2C flight control systems (172 D,E); Brittain Industries, P.O. Box 51370, Tulsa, OK 74151.

SA2256WE: Installation of leading edge cuff on each wing; Marshall E. Quackenbush, P.O. Box 2421, California City, CA 93505.

SA2281CE: Installation of aileron and/or flap gap seals; Horton STOL Craft, Inc., Wellington Muni. Apt., Wellington, KS 67152.

SA2375SW: Install Lycoming O-320-D2G engine rated for 160 hp takeoff and 150 METO (172 D,E,F,G,H); RAM Aircraft Modifications, Inc., P.O. Box 5219, Madison Cooper Airport, Waco, TX 76708.

SA2517CE: Installation of replacement wingtips (most 172); Horton STOL Craft, Inc., Wellington Muni. Apt., Wellington, KS 67152.

SA2518CE: Installation of replacement wingtips (172 E-J); Horton STOL Craft, Inc., Wellington Muni. Apt., Wellington, KS 67152.

SA2551CE: Establishes the compatibility to install the flap actuated droop system; Horton STOL Craft, Inc., Wellington Muni. Apt., Wellington, KS 67152.

SA2651NM: Installation of Lycoming O-360-A3A or A4A engine and McCauley propeller (172 N,P); Lee R. Smith, 224 Cougar Flat Rd., Castle Rock, WA 98611.

SA2671WE: Installation of upper wing surface stall fences and aileron gap sealing strips (172 M); Group One, 2915 Earhat Apron, Torrance, CA 90505.

SA2852: Installation of leading edge cuffs, stall fences, aileron seals, wingtips (172, 172 A,B,C,D,E,F,G,H,I,K,L,M); Bush Conversions, P.O. Box 431, Udall, KS 67146.

SA293NW: Installation of Franklin 6A0350 engine and McCauley 2A31C21/84S06 propeller; Seaplane Flying, Inc., 1111 S.E. 5th Street, Vancouver, WA 98661.

SA296GL: Convert to conventional landing gear (172, 172 A,B,C,D,E,F,G,H,I,K,L,M,N); Robert L. or Barbara V. Williams, Box 608, Udall, KS 67146.

SA3-576: Installation of speed kit (172, 172 A,B) Aircraft Development Co., 1326 North Westlink Blvd., Wichita, KS 67212.

SA3-107: Rudder trim system; Javelin Aircraft Co., Inc., 9175 East Douglas, Wichita, KS 67207.

SA3-571: Lycoming O-360-A1D engine, McCauley 2D36C14-78KM-4 or Hartzell HC-C2YK-1A/7666-2 propellers (172,172 A,B,C,D,E,F,G,H); KWAD Co., 4530 Jettridge Dr., NW, Atlanta, GA 30327.

SA3-126: Lycoming C-340-A1A engine and Hartzell HC-82XG-1DB/8433-12 propeller; KWAD Co., 4530 Jettridge Dr., NW., Atlanta, GA 30327.

SA3-690: Installation of Doyn swept-tail assembly kit 1400; KWAD Co., 4530 Jettridge Dr., NW, Atlanta, GA 30327.

SA3-655: Wheel fairings; Marsh Flying Club, Inc., 812 East 34th St., Sioux Falls, SD 57101.

SA3-333: Installation of Doyn fiberglass wheel fairings; KWAD Company, 4530 Jettridge Dr., NW, Atlanta, GA 30327.

SA3105WE: Installation of flap gap fairing (172,172 A,B,C,D,E,F,G,H,I,K,L,M,N normal and utility categories); Thermal Aircraft Company, 56-850 Tyler Street, Thermal, CA 92274.

SA319GL: Installation of Bolen wheel extenders on Cessna models modified per STC SA296GL (172 D,E,F,G,H,I,K); Robert L. or Barbara V. Williams, Box 608, Udall, KS 67146.

SA332GL: Installation of Lycoming O-360 Series engine and McCauley 1A200/DFA propeller (172 I,K,L,M,N,P seaplane); Penn Yan Aero Service, Inc., 2499 Bath Rd., Penn Yan, NY 14527.

SA3788WE: Installation of the "liftrol" wing spoiler system (172, 172 A,B,C,D,E,F,G,H, I,K,L); Esther E. Cleaves, 4542 Norma Drive, San Diego, CA 92115.

SA383GL: Convert to conventional landing gear (R172 K); Robert or Barbara Williams, Box 608, Udall, KS 67146.

SA389NE: Installation of AVCO Lycoming O-320-E2D modified (172 M); Raymond R. Cloutier, 112 Suffolk St., Providence, RI 02908.

SA4-893: Installation of wheel fenders on nose gear and main gear (172); Aurora Air, Inc., DBA Madras Air Service, Madris, OR 97741.

SA4080SW: Installation of leading edge cuffs, stall fences, aileron seals, wingtips (172 M,N); Bob Williams, DBA Bush Conversions, P.O. Box 431, Udall, KS 67146.

SA4120SW: Add sheet metal flap seals and/or flexible fabric aileron gap seals; Flight Bonus, Inc., P.O. Box 120773, Arlington, TX 76012.

SA414NW: Installation of Lycoming O-360-A1F6D engine and McCauley Propeller (172 H); Salem Aviation, 2680 Aerial Way, S.E., Salem, OR 97302.

SA420CE: Lycoming O-360-A1A or A1D engine, McCauley 2D36C14/78KM-4 or Hartzell KC-C2YK-1B/7666-2 propeller (172, 172 A,B,C,D,E,F,G,H landplanes; 172 D,E,F,G,H floatplanes; EDO 89-2000 floats only); Robert L. or Barbara V. Williams, Box 431, 213 North Clark, Udall, KS 67146.

SA4428SW: Installation of Lycoming O-360-A2F engine and McCauley 1A170/CFA-76-60 propeller (172 I,K,L,M,N); Air Plains Svcs. Corp., Box 541, Wellington, KS 67152.

SA44SO: Cessna L-19 wing conversion; Marrs Aircraft, Avion Park, FL 33825.

SA480CE: Nose and main gear wheel fairing (Jetstreams) using standing unaltered Cessna landing gear and nose gear for model and year using 500 x 5 nose and 600 x 6 main gear tires (172); Creative Designs, 1338 Orkla Dr., Minneapolis, MN 55427.

SA499NW: Installation of Robertson Aircraft STOL kit with Robertson leading edge cuff, fence, Cessna wing flap filler panel; Seaplane Flying, Inc., 1111 S.E. 5th Street, Vancouver, WA 98661.

SA5433SW: Conversion from tricycle to conventional landing gear (172, 172 A,B,C,D,E,F,G,H,I,K,L,M,N,P); Aircraft Conversion Technologies, Inc., 1410 Flightline Dr., Hangar A, Lincoln Airport, Lincoln, CA 95648.

SA564NE: Installation of the Textron Lycoming O-360-A4A, A4M, A4N engine and Sensenich Model 76EM8S14-060 propeller; Alpine Aircraft Inc., Route 12, Box 94, Oaks Road, Hagerstown, MD 21740.

SA596WE: Installation of Doyn Aircraft Co. kit per STC SA3-571 (172 D,E (seaplane); Barbara or Bob Williams, Box 431, 213 North Clark, Udall, KS 67146.

SA610SW: Lycoming O-320 and O-360 series engines with Hartzell HC-82XL-7636D-4; HC-C2YK-2R/7666 or HC-920Z-2B/8447-128 propeller (172); R.F.B. Company, Chandler Field, Purcell, OK 73080.

SA647CE: Install Lycoming O-360-A1A 180-hp engine and constant-speed propeller (172 I,K,L,M,N land and floatplanes using EDO 89-2000 floats only); Barbara or Bob Williams, Box 431, 213 North Clark, Udall, KS 67146.

SA672NE: Installation of fiberglass reinforced plastic nose gear wheel fairing (172 through Q); Maple Leaf Avn. Ltd., Box 160, Brandon, Manitoba, R7A5Y8 Canada.

SA672SW: Cessna 305 (L-19) wings; James A. Summersett, Jr., 418 Redcliff Dr., San Antonio, TX 78216.

SA703GL: Installation of Lycoming O-360-A4A, A4M, or A4N series engines and Sensenich propeller models 76EM8S5 or 76EM8SPY (172 I,K,L,M,N,P); Penn Yan Aero Service Inc., 2499 Bath Road, Penn Yan, NY 14527.

SA765WE: SP51463-1A speedplates; Grand Valley Acft., Inc., P.O. Box 158, Grand Junction, CO 81501.

SA792WE: Lycoming O-360 series engines and Sensenich M67EMM propellers (some 172s); Aircraft Conversion Technologies, Inc., 1410 Flightline Dr., Lincoln, CA 95648.

SA807CE: Installation of Lycoming 180-hp engine and constant-speed propeller (172, 172 A through I,K,L,M,N; landplane, normal category only); Robert L. or Barbara V. Williams, 117 East First, Udall, KS 67146.

SA81WE: Aileron and rudder trim controls; Met-Co-Aire, P.O. Box 2216, Fullerton, CA 92633.

SA87NW: Installation of Lycoming O-360-A1A 180-hp engine and Hartzell HC-C2YK-1A/7666A-O propeller (172 I,K,L); Kenmore Air Harbor, Inc., P.O. Box 64, Kenmore, WA 98028.

SA902CE: Installation of STOL kit (172, 172 A through 172 I, 172 K, 172 L, 172 M, 172 N, 172 P); Robert L. or Barbara V. Williams, Box 608, Udall, KS 67146.

SA910CE: Installation wing leading edge cuffs, drooped tips, stall fences and aileron gap seals (172 landplane, 172 A through 172 P landplane and floatplane); Horton STOL-Craft, Inc., Wellington Municipal Airport, Wellington, KS 67152.

SA995SW: Increase wing leading edge radius wingtip fairings, and landing light lens; Bob or Barbara Williams, Box 608, Udall, KS 67146.

FLOATS AND SKIS

SA1000EA: Installation of Pee Kay Model 2300 or B2300 (TSO-C27-2) seaplane floats (172 I,K,L,M,N seaplane); Pee Kay Floats DeVore Aviation Corporation, 6104-B Kircher Boulevard, N.E., Albuquerque, NM 87109.

SA1168SO: Installation of Aqua Model 2200 floats (172 N); Claggett Aircraft Products Research and Engineering, Inc., 805 Geiger Rd., Zephyrhills, FL 33599.

SA340NW: Installation of Pee Kay Model 2300 floats (172, 172 D,K); Seaplane Flying, Inc., P.O. Box 2164, Vancouver, WA 98661.

SA358NW: Installation of Pee Kay Model 2300 seaplane floats (172 A through 172 N); DeVore Aviation Corporation, 6104-B Kircher Blvd., N.E., Albuquerque, NM 87109.

SA365NW: Installation of Pee Kay Model 2300 seaplane floats. Approval is intended for installation in listed models with Franklin 6A-0360-C2 engine installation IAW Seaplane Flying, Inc., STC SA340NW (172 D through 172 K); DeVore Aviation Corp., 6104-B Kircher Blvd., N.E., Albuquerque, NM 87109.

SA460NW: Installation of Pee Kay Model 2300 or B2300 seaplane floats (R172 K); De-Vore Aviation Corp., 6104-B Kircher Blvd, NE, Albuquerque, NM 87109.

SA584NW: Installation of EDO model 689-2130 seaplane floats (172 L,M,N,P); EDO Corporation Government Systems Division, Farmingdale OPS, Republic Airport, East Farmingdale, NY 11735.

SA650CE: Installation of Fleet Model 2500 floats (172 A through 172 K); Michael Kelsey, 22315 N. Forest Loop Rd., P.O. Box 153, Granite Falls, WA 98252.

SA665CE: Installation of model 2400 Aqua floats (172 D,E,F,G,H,I,K,M,N,R172 K); Claggett Aircraft Products Research and Engineering, Inc., 805 Geiger Rd., Zephyrhills, FL 33599.

SA3-179: A3500A or A2500A main skis and NA1200A nose skis; FluiDyne Engineering Corp., 5900 Olson Memorial HWY, Minneapolis, MN 55422.

SA3-256 Main ski AWB2500A and nosewheel ski AWN1200; FluiDyne Engineering Corp., 5900 Olson Memorial HWY, Minneapolis, MN 55422.

SA3-355: C3000 main ski and AWN1200 nosewheel ski; FluiDyne Engineering Corp., 5900 Olson Memorial HWY, Minneapolis, MN 55422.

SA3-614: Model 3000 main skis and model 2000 nose ski; FluiDyne Engineering Corp., 5900 Olson Memorial HWY, Minneapolis, MN 55422.

SA315GL: Installation of FluiDyne skis (172, 172 A,B,C,D,E,F,G,H,I,K,L,M,N); Robert L. or Barbara V. Williams, Box 608, Udall, KS 67146.

SA88CE: C-3000 main skis and AWN-1200A nose ski; FluiDyne Engineering Corp., 5900 Olson Memorial HWY, Minneapolis, MN 55422.

BRAKE SYSTEMS

SA119NE: Replace original brake discs with discs modified in accordance with Roland Langarzo procedure; Orel Aviation, Inc., 218 Hollywood Ave., Valley Stream, NY 11581.

SA1311CE: Chrome-plated brake disc installation (172 F,G,H,I,K,L); Engineering Plating and Processing, Inc., 641 Southwest Boulevard, Kansas City, KS 66103.

SA13GL: Installation of Cleveland wheel No. 40-97A and brake No. 30-63A; Aircraft Wheel & Brake Div., Parker Hannifin Corp., 1160 Center Rd., Avon, OH 44011.

SA1584SO: Installation of Appalachian Accessories brake rotor P/N 75-27; Appalachian Accessories, P.O. Box 1077, TCAS, Blountville, TN 37617.

SA1760SO: Installation of Appalachian Accessories stainless steel brake rotors P/N 75-15C; Appalachian Accessories, Tri-City Airport Station, Blountville, TN 37617.

SA1761SO: Installation of Appalachian Accessories stainless steel brake rotors P/N 75-27; Appalachian Accessories, Tri-City Airport Station, Blountville, TN 37617.

SA1979SO: Installation of Appalachian Accessories stainless brake discs (172 L); Appalachian Accessories, P.O. Box 1077, Tri-City Airport Station, Blountville, TN 37617.

SA2189SO: Replacement of Cleveland brake discs with Appalachian Accessories stainless steel brake discs; Appalachian Accessories, Tri-City Airport Station, Blountville, TN 37617.

SA2190SO: Replacement of Cleveland brake discs with Appalachian Accessories stainless steel brake discs; Appalachian Accessories, Tri-City Airport Station, Blountville, TN 37617.

SA2297SO: Replacement of Cleveland brake discs with Appalachian Accessories stainless steel brake discs; Appalachian Accessories, P.O. Box 1077, Blountville, TN 37617.

SA2381SO: Installation of Appalachian Accessories stainless steel brake rotors P/N 75-92; Appalachian Accessories, Tri-City Airport Station, Blountville, TN 37617.

SA2382SO: Installation of Appalachian Accessories stainless steel brake rotors P/N 75-93; Appalachian Accessories, Tri-City Airport Station, Blountville, TN 37617.

SA4-394: Installation of George M. Adams DGA-2000 wheels and Bodell Manufacturing Model BMI-252 brakes; Malcom E. Bodell, 23804 Penn Ave., Torrance, CA 90501.

SA4704NM: Installation of an anti-skid pressure accumulator module for the hydraulic brake system; Long's Aircraft Service, 6331 South C St., Tacoma, WA 98408.

FUEL SYSTEMS

SA1124GL: Installation of Shadin Co. fuel flow indicating system; Shadin Co, Inc., 6950 Wayzata Blvd., Minneapolis, MN 55426.

SA124EA: Shur-Vent fuel tank filler cap; Alfred R. Puccinelli, DER 1-145, 36 Bay Dr., East, Huntington, LI, NY 11743.

SA1395SO: Remove pipe plug, Cessna P/N MS20913-1 or MS20913-1D from bottom of fuel selector valve body and replace with drain valve, MT-1072-77 (172 I,K,L,M); C-Mods, Inc., P.O. Box 15388, Durham, NC 27704.

SA1396SO: Remove drain plug, Cessna P/N AN814-3DL from bottom of fuel selector valve body and replace with drain valve MT-1072-77 (172 N); C-MODS, Inc., P.O. Box 15388, Durham, NC 27704.

SA1614WE: Installation of one 12-gallon auxiliary fuel tank in each outboard wing panel; Flint Aero, Inc., Div. of Oronco, Inc., P.O. Box 1458, Spring Valley, CA 92077.

SA1948CE: Operation on unleaded and/or leaded automotive gasoline (172 I,K,L,M); Petersen Aviation, Inc., Route 1, Box 19, Minden, NE 68959.

SA2006CE: Use of leaded and unleaded automotive fuel; Petersen Aviation, Inc., Route 1, Box 18, Minden, NE 68959.

SA2600CE: Operation of airplane on leaded and unleaded automotive gasoline (172 I,K,L,M,N,P); Petersen Aviation, Inc., Route 1, Box 18, Minden, NE 68959.

SA2601CE: Operation of airplane on leaded and unleaded automotive gasoline (172 I,K,L,M,N,P 160hp); Petersen Aviation, Inc., Route 1, Box 18, Minden, NE 68959.

SA326GL: Installation of fuel tank caps (172, 172 A,B,C,D,E,F,G,H,I,K,L,M); John J. Francissen, 426 Pleasant Dr., Roselle, IL 60172.

SA3585NM: Installation of ARNAV System FC-10 or FT-10 digital fuel flow indicating system (R172 K); ARNAV Systems, Inc., 26810 Oak Ave., Suite J., Canyon Country, CA 91351.

SA4-198: 13-gallon auxiliary fuel system; Met-Co-Aire, P.O. Box 2216, Fullerton, CA 92633.

SA4048WE: Installation of Silver Instruments fuel Fueltronic IG or Fueltronic IP or Fuelguard GRD digital fuel flow and totalizer system (R172 K); Silver Instruments, Inc., 2346 Stanwell Dr., Concord, CA 94520.

SA4242WE: Installation of SDI CFS-1000, -1001, of FT-100 fuel flow indication system and facet P/N 480543 auxiliary fuel pump (172, 172 A through 172 H (landplane only, normal category); Del-Air, P.O. Box 746, Strathmore, CA 93267.

SA615NE: Installation of an 18-gallon fuel tank in the aft baggage compartment; O & N Aircraft Modification, Inc., P.O. Box 292, Seamans Airport, Factoryville, PA 18419.

SA693GL: Use of unleaded automotive gasoline; EAA, Wittman Airfield, Oshkosh, WI 54903.

SA703NE: Installation of Sheltech SF500 or SF510 fuel data computer; Gerard Bolies, Box 1038, Hangar H East, Hampton Airport, Wainscott, NY 11925.

SA733WE: 12.7-gallon auxiliary fuel tank; Aircraft Modifications and Repair, 20331 Hesperian Blvd., Hayward, CA 94541.

SA761GL: Modify airplane to fly on unleaded automotive gasoline, 87 minimum anti-knock index (172, 172 A,B,C,D,E,F [T-41A], G,H [T-41A]); EAA, P.O. Box 2592, Oshkosh, WI 54903.

SA801GL: Modify airplane to fly on unleaded automotive gasoline 87 minimum anti-knock index (172 I,K,L,M); EAA, Wittman Airfield, Oshkosh, WI 54903.

SA890SW: Medairco Automatic Fuel Alert F-2; Medairco, 3601 East Admiral Pl., Tulsa, OK 74150.

EXTERIOR LIGHTING

SA21GL: Installation of Grimes multi-plexed anti-collision light system (172, 172 A through 172 I,K,L,M,N,P); Grimes Manufacturing Co., 515 North Russell St., Urbana, OH 43078.

SA242EA: 40-0040 and 40-0043 supplemental light system; Grimes Manufacturing Co., 515 North Russell St., Urbana, OH 43078.

SA3-89: Equipment shelves and Grimes D-7080 rotating beacon; Capitol Aviation, Inc., Capitol Airport, Springfield, IL 62705.

SA3-75: Grimes rotating anti-collision light; Mid States Aviation Corp., Sky Harbor Airport, Northbrook, IL 60062.

SA3338WE: Recognition light installation on horizontal stabilizer (172, 172 A,B,C,D,E,F [T-41], H [T-41], I,K,L,M,N); DeVore Aviation Corporation, Suite B, 6104 Kircher Blvd., N.E., Albuquerque, NM 87109.

SA3418WE: Installation of SDI Novastar II combined position/anticollision lighting system; Symbolic Displays, Inc., 1762 McGaw Avenue, Irvine, CA 92714.

SA3487WE: Installation of tail flood light (P172 D); DeVore Aviation Corp., Suite D, 6104 Kircher Blvd., NE, Albuquerque, NM 87109.

SA4-306: Roto-Nav bracket and Grimes D 7080 anti-collision light on vertical fin (172); Air Oasis Company, Municipal Airport, Long Beach, CA 90801.

SA4005NM: Installation of Precise Pulselite control unit on the landing/taxi system; Precise Flight, Inc., 63120 Powell Butte Rd., Bend, OR 97701.

SA4337NM: Installation of landing/recognition lights; R.M.D. Aircraft Lighting, Inc., 3648 Roanoke Ct., Hillsboro, OR 97123.

SA43EA: Installation of Re-Vue Agency nite-lite and electronic aerial advertising sign (172, 172 A,B,C); Re-Vue Agency, Inc., 616 East 4th. St., Pueblo, CO 80204.

SA4711NM: Installation of Sunrise Aviation Sunliter landing light; PMDB, Inc., DBA Sunrise Aviation, 1314 26th Ave., NW, Gig Harbor, WA 98335.

SA533CE: Installation of Canairco 1214WS supplementary light; Canairco Ltd., 400 First Avenue North, Minneapolis, MN 55401.

SA615EA: Installation of Whelen anti-collision strobe light system, models H through D, HR or HS (-14 (14V) or -28 (28V)), as replacement for originally installed anti-collision lights; Whelen Engineering Co., Inc., Winter Ave., Deep River, CT 06417.

SA64CE: Sky-Lite B-100 beacon light; Gas Equipment and Engine Co., 1015 Diamond Ave., Evansville, IN 47708.

SA6NE: Installation of Whelen anti-collision strobe light power supplies installed IAW STC SA615EA or SA800EA (172 M): Whelen Engineering Co., Inc., Winter Ave., Deep River, CT 06417.

SA800EA: Installation of Whelen anti-collision strobe light system, various models; Whelen Engineering Co., Inc., Winter Ave., Deep River, CT 06417.

SA917EZ: Installation of Grimes Manufacturing Co. aviation white anti-collision strobe light systems two light or three light series 555 for wingtips and tail; Grimes Division/ FL Aerospace Corp., 515 W. Russell St., Urbana, OH 43078.

SAFETY

SA1196GL: Install SAF-T-STOP auxiliary seat stop device (most 172); Glen A. Florence, Aero Technologies, Inc., 39745 Sylvia, Mt. Clements, MI 48045.

SA1210GL: Install SAF-T-STOP auxiliary seat stop device (R172 E-K); Glen A. Florence, Aero Technologies, Inc., 39745 Sylvia, Mt. Clements, MI 48045.

SA2067NM: Installation of a TSO shoulder harness and BAS, Inc. harness system; Bud's Aero Specialties, Inc., P.O. Box 190, Eatonville, WA 98328.

SA2960NM: Install "stay put" seat position locking device; B&D Safety Lok Co., 14409 141st Ave, SE, Renton, WA 98056.

SA3643SW: Installation of seat restraint belt, pilot and/or copilot; The Rebound Co., P.O. Box 656, 510 Ave. E, Marble Falls, TX 78654.

SA825GL: Installation of safety belt extenders; John Stevenson, 8637 Huckleberry Lane, Lansing, MI 48917.

SA861WE: Pilot and passenger restraint harness; Pacific Scientific Co., 1346 South State College Blvd., Anaheim, CA 92803.

SA864CE: Installation of electro pneumatic stall warning system (172, S/N 17254893 and on); Keaton Engineering Company, 1000 West 55th Street South, Wichita, KS 67217.

SA93EA: Installation of Monitair angle of attack system; Rosemount Engineering Co., 12001 West 78th St., Eden Prairie, MN 55343.

MISCELLANEOUS

SA1-193: Conversion of aircraft for parachute jumping operations (172); Parachute Club of America, P.O. Box 409, Monterey, CA 93940.

SA1-426: Auxiliary seat in baggage compartment; William A. Welch, Boyce Road, RD-1, Danbury, CT 06810.

SA1-545: External electrical power receptacle; R.H. Haynes, Inc., 62 Voorhis Lane, Hackensack, NJ 07601.

SA105NW: Mounting and wiring for occasional installation and use of Barnes Engineering Company airborne fire spotter system (172); State of Washington, Dept. of Natural Resources, Route 13, Box 62, Olympia, WA 98501.

SA115NW: Modification of right side door and the addition of a jump seat; Charles P. Bunch, 1211 Donovan Ave., Bellingham, WA 98225.

SA1211EA: Installation of EPA portable enviro-pod with Tensor-Hasselbad camera configuration (172 M,N); Tensor Industries, Inc., 117 Schley Avenue, Lewes, DE 19958.

SA1258EA: Installation of Cosco Model 78 child restraint system (172 M); Stuart R. Miller, P.O. Box 926, Grand Central Station, New York, NY 10163.

SA146GL: Installation of EPA portable enviro pod (172 M,N); Air Force Wright Aero Lab, AFWAL/AARF.

SA1513SO: Installation of front seat modification; Civil Air Patrol, Maxwell AFB, AL 36112.

SA152NW: Installation of deck and glare shield; R.J. Schoers, 4479 139th Ave., SE, Bellview, WA 98006.

SA1531NM: Modification of the Union Aviation Supplemental Type Certification No. SA860SO for portable rudder/brake hand control by adapting the control to Nylafil rudder pedals (172 I,K,L,M,N,P); Dave Flock, 1009 West 31st Street, Loveland, CO 80537.

SA1813NM: Installation of Lompoc Aero inner plexiglass pane in swing-out window frames presently employing a single window (172 A,B,C,D,E,F,G,H,I,K,L,M,N, P,T-41A); Lompoc Aero Specialties, 2772 North Rancho Dr., Las Vegas, NV 89130.

SA1824NM: Installation of Lompoc Aero inner plexiglass pane and swing out window frames presently employing a single window (172 RG); Lompoc Aero Specialties, 2772 N. Rancho Dr., Las Vegas, NV 89130.

SA1826CE: Installation of photo bubble in baggage door opening (172 I,K,L,M,N,P); Hue Aire, Inc., 930 North Main, Hutchinson, KS 67501.

SA1827CE: Installation of photo bubble in baggage door opening (172 RG); Hue Aire, Inc., 930 North Main, Hutchinson, KS 67501.

SA2-1178: Replacement of right front seat with stretcher; Aviation Fabricators, 504 Price Lane, Clinton, MO 64735.

SA2-1533: Installation of windshield wiper; Curtis-Wright/Marquette, Inc., 400 South Main St., Fountain Inn, SC 29644.

SA211GL: Installation of windshield P/N W-2055, W/T-2055, W/G-2055 on Cessna 172 and 172 A, and P/N W-2056, W/T-2056 or W/G-2056 on Cessna 172 B and 172 C; Great Lakes Aero Products, 1021 North Chevrolet Ave., Flint, MI 48504.

SA21NW: Installation of 50 lb. baggage compartment in EDO floats; Kenmore Air Harbor, Inc., P.O. Box 64, Kenmore, WA 98028.

SA220SO: Tow bar, release mechanism, and safety tow link; Gasser Banners, Inc., P.O. Box 3502, Metropolitan Airport, Nashville, TN 37217.

SA2362WE: Installation of lure toxicant dispenser AEY 70, Model # 1 (172); USDA, 3701 West Nob Hill Blvd, Yakima, WA 98902.

SA238SW: Installation of cargo door; Ashley Hawk Aviation, Stinson Municipal Airport, 8505 Mission Road, San Antonio, TX 78214.

SA2575WE: Installation of observer station in rear seat-baggage compartment area, searchlight and dual speakers in bottom of aft fuselage, windows in baggage compartment area (baggage was removed), and a muffler (172 K,M, normal category landplane only); World Associates, Inc., 430 Wilshire Boulevard, Suite 204, Santa Monica, CA 90401.

SA3-82: Retractable lifting handles model H,P/N 50-9; Woychik Aircraft Equipment, Middleton, WI 53562.

SA3751NM: Installation of baggage door; Del-Air, 2121 S. Wildcat Way, Porterville, CA 93257.

SA3798NM: Installation of one piece windshield; Del-Air, P.O. Box 746, Strathmore, CA 93267.

SA4-1601: Conversion of aircraft for parachute jumping and aerial photography operations (172); Air Oasis Company, Municipal Airport, Long Beach, CA 90801.

SA4-620: Installation of baggage door on left side of fuselage; The Air Oasis, P.O. Box 1769, Long Beach, CA 90801.

SA4-662: Camera pod; Clifton G. McBride, 620 Rollingwood Dr., Vallejo, CA 94590.

SA414SW: Jump step or mud guard on right landing gear; Curtis J. Smith Jr., Route 2, Box 314, Breaux Bridge, LA 70517.

SA4283WE: Installation of electric inflatable door seal (172, 172 A,B,C,D,E,F,G,H,K, L,M,N); Bob Fields Accessories, 5673 Stanford Street, Ventura, CA 93003.

SA462WE: Jump door 62-500 (172); Sky Motive, Inc., Route 1, Box 32, Snohomish, WA 98290.

SA4721NM: Installation of Syncon, Inc. Alarm System, Model 100C, on non-pressurized metal structure aircraft; Syncon, Inc., P.O. Box 188, Loon Lake, WA 99148.

SA5-12: Rear canvas hammock seat; Aircraft Rebuilders, 1803 Lake Otis Rd., Anchorage, AK 99500.

SA531NW: Installation of windshield; Jack Shannon, 215 Terminal Building, Boeing Field, Seattle, WA 98101.

SA548GL: Installation of Flitefile console (172 1965 through 1981); John Stevenson, 8637 Huckleberry Lane, Lansing, MI 48917.

SA7362SW: Installation of camera provisions (172 P); Airmotive Engineering, Rte 1, Box 468, New Braunfels, TX 78130.

SA7398SW: Modified door latch; John Benham, Rt 2, P.O. Box 6950, Pipe Creek, TX 78063.

SA895WE: Kienzla 65 flight time recorder (172); Burrows and Sons, 1829 Bridgeport Avenue, Claremont, CA 91711.

SA933NW: Installation of Blancher Model 101 stowable seats in place of second row seats; BAS, P.O. Box 190, Eatonville, WA 98328.

SA94CE: Removal of right cabin door; Skylane Flying Service, Inc., 2029 Allens La., Evansville, IN 47708.

SA981WE: Two detachable folding rear seats replacing original seats; Missionary Aviation Fellowship, P.O. Box 32, Fullerton, CA 92632.

SA992CE: Installation of 10-inch diameter camera hole (172 I,K,L,M,N,P); Ryan International Airways Corporation, 1640 Airport Rd., Mid-Continental Airport, Wichita, KS 67209.

POPULAR MODIFICATIONS

The following are some of the more popular modifications owners can make to a Cessna 172.

STOL

STOL conversions are perhaps king of all the modifications available to the 172 owner. STOL is the military designation for short take-off and landing aircraft. STOL has been extended into general aviation markets, resulting in spectacularly performing conversion aircraft. The typical STOL modification involves changes to the overall shape of the wing (usually in the form of a leading edge cuff), the addition of stall fences (to stop a stall from progressing along the wing spanwise (Fig. 11-1)), gap seals, wingtips, and perhaps an increase in power. Horton STOLcraft specifications for a converted Cessna 172 (maximum performance) are:

- Gross weight: 2200 lbs
- Takeoff Speed: 38 mph
- Takeoff over 50' obst: 840 ft
- Cruise Speed: 133 mph
- Approach Speed: 38 mph
- Landing over 50' obst: 625 ft

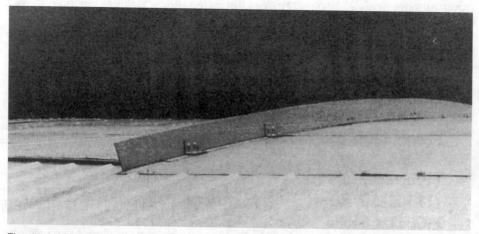

Fig. 11-1. Most STOL conversions make use of stall fences to inhibit the progression of a stall along the wing's surface.

Sometimes an owner will make STOL modifications one part at a time, and often with STCs from several sources. Before proceeding with this method, check with your local FSDO about the various STCs you are considering because mixed STCs might not be compatible. STOL modifications and kits are available from:

AVCON Conversions, Inc.
P.O. Box 654
Udall, KS 67146
(316) 782-3317

Bush Conversions
P.O. Box 431
Udall, KS 67146
(800) 752-0748
(316) 782-3851

Horton STOLcraft
Wellington Municipal Airport
Wellington, KS 67152
(800) 835-2051
(316) 326-2241

Sierra Industries
P.O. Box 3184
Uvalde, TX 78802
(512) 278-4381

Power

Power modifications are the second most popular type of work done to 172s, and are often done in conjunction with STOL modifications. These mods consist of engine replacement (Fig. 11-2) to increase the useful load and flight performance figures of the aircraft. Power modifications can be extensive and costly, although most are not much above the level of a good engine rebuild charge. The following firms provide power modifications for Cessna 172 airplanes:

Avcon Industries, Inc., installs the Lycoming O-360-A1A engine with a constant-speed propeller on the 172. For further information, contact:

AVCON Conversions, Inc.
P.O. Box 654
Udall, KS 67146
(316) 782-3317

Bush Conversions
P.O. Box 431
Udall, KS 67146
(800) 752-0748
(316) 782-3851

Penn Yan Aero installs the 180-hp O-360 engine in the 172 and claims that it increases the cruise speed from 139 to 160 mph. The rate of climb is brought from 730 fpm to 1100 fpm and the takeoff distance is reduced from 730 to 400 feet while retain-

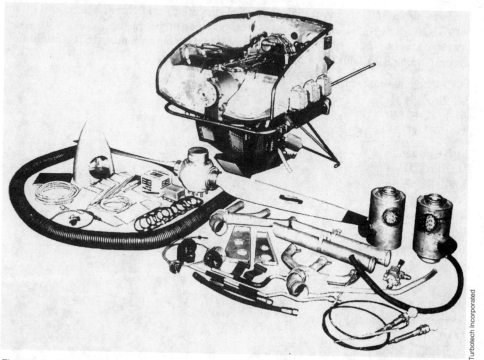

Fig. 11-2. Typical parts kit for an engine conversion including mount, baffles, exhaust, harnesses, and the like.

ing the simplicity of a fixed-pitch propeller. Penn Yan refers to their modification as the Superhawk (Fig. 11-3). For further information, contact:

Penn Yan Aero
Penn Yan Airport
2499 Bath Rd.
Penn Yan, NY 14527
(315) 536-2333

RAM Aircraft Corp. installs the Avco Lycoming O-320-D2G 160-hp engine in all model 172s to replace the older Continental O-300 145-hp engines and to replace AD-plagued O-320-H2AD. For more information, contact:

RAM Aircraft Corp.
Waco-Madison Cooper Airport
P.O. Box 5219
Waco, TX 76708
(817) 752-8381

Aero Mods of Arizona can install the Franklin 220-hp engine or a turbocharged Franklin 250-hp engine on 172 and 175 models. The modifications boost maximum speed to 170 mph and rate of climb to 1800 fpm. This particular conversion is popular among the floatplane fliers, giving a cruise of 155 mph (with floats).

Fig. 11-3. The 180-hp Superhawk by Penn Yan Aero.

The Franklin engine makes this the most powerful conversion available for the 172/175 series, turning the airplane into a real four-seat powerhouse, perfect for getting out of tight places with heavy loads. Flying a 220- or 250-hp 172 is a real thrill and will give you a look at the hidden capabilities of this fine airplane.

For further information, contact:

Aero Mods of Arizona
P.O. Box 1010
Sonoita, AZ 85637
(602) 455-9273

Wingtips

One of the most popular modifications for 172s is wingtip changes to increase performance (Figs. 11-4 and 11-5). Sighard Hoerner, Ph.D, designed a high-performance wingtip for the U.S. Navy that eventually led to development of improved wingtips for lightplanes. A properly designed wingtip can provide an increase of 3–5 mph in cruise speed and a small increase in climb performance, but most important are the improved low-speed handling characteristics: 10—20 percent reduction in takeoff roll, 4–5 mph lower stall speed, and improved slow-flight handling. Installation time can be as few as two or three hours. For more information, contact:

Ace Deemers
Madras Air Service
1914 NW Deemers Dr.
Madras, OR 97741

Met-Co-Aire
P.O. Box 2216
Fullerton, CA 92633
(714) 870-4610

Fig. 11-4. The downturned style wingtip is often seen on Cessna airplanes and is available from Ace Deemers.

Met-Co-Aire

Fig. 11-5. The Met-Co-Aire (Hoerner-style) wingtip.

Landing gear

Taildragger conversions have become popular among owners of 172s. The nose-wheel is removed, the main gear moved forward, and a tailwheel installed (Fig. 11-6). Performance benefits of 8–10 mph increase in cruise speed, shorter takeoff distances, and better rough-field handling are claimed.

If a rough-field aircraft is the owner's ultimate goal, then STOL and increased power modifications should be made simultaneously. Because the multiple modifications would cost so much, a 172 owner might consider purchasing a more appropriate aircraft, such as a Cessna 180 or 185—or if you wanted to go outside the Cessna line, a Maule.

Taildragger modifications and kits are available from:

Aircraft Conversion Technologies, Inc.
1410 Flightline Dr., Hangar A, Lincoln Airport
Lincoln, CA 95648
(916) 645-3264

Fig. 11-6. Taildragger conversion.

Bush Conversions, Inc.
P.O. Box 431
Udall, KS 67146
(800) 752-0748
(316) 782-3851

Ron Fravel
4465 Highway 3375
Corydon, IN 47112
(812) 738-1902

Gap seals

Gap seals are extensions of the lower wing surface from the rear spar to the leading edge of the flap or aileron or both. The seals cover approximately six square feet of open space, allowing a smoother flow of air around the wing. In addition to the reduction of parasitic drag, modifiers claim that the aircraft will cruise 1–3 mph faster and stall 5–8 mph slower. Gap seals are often part of a STOL conversion (Fig. 11-7).

Gap seals can be made of sheet metal or a flexible strip of fiberglass. The strips are riveted into place and generally require two or three hours of installation time. For further information, contact:

B&M Flap Seals
P.O. Box 431
Udall, KS 67146
(800) 752-0748
(316) 782-3851

Horton STOLcraft
Wellington Municipal Airport
Wellington, KS 67152
(800) 835-2051
(316) 326-2241

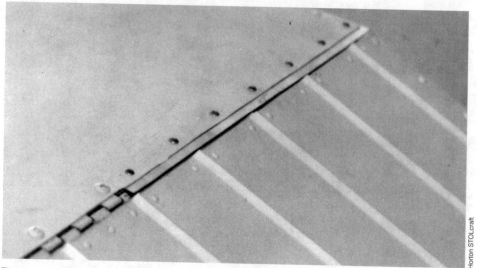

Fig. 11-7. Aileron gap seals provide a smooth flow of air over the wings.

Les Leonard
P.O. Box 32
Aguila, AZ 85321
(602) 685-2637

Weight

STCs for 172N and P models increase the gross weight to 2550 pounds. For further information, contact:

Air Plains Svcs. Corp.
P.O. Box 541
Wellington Airport
Wellington, KS 67152
(316) 326-8904

Penn Yan Aero
Penn Yan Airport
2499 Bath Rd.
Penn Yan, NY 14527
(315) 536-2333

Fuel

Larger or additional fuel tanks are usually installed to increase the airplane's operational range or to accommodate a high-performance engine that consumes fuel faster than a stock engine. Fuel mods require proper paperwork to be legal (the STC) plus supportive documentation if the additional fuel capacity is questioned by a law enforcement agency; check the FARs and ask the conversion vendor for up-to-date specific information about documentation requirements. Information about optional fuel tanks is available from:

Aircraft Conversion Technologies, Inc.
1410 Flightline Dr., Hangar A, Lincoln Airport
Lincoln, CA 95648
(916) 645-3264

Flint Aero, Inc.
1935 N. Marshall Ave.
El Cajon, CA 92020
(619) 448-1551

Noise reduction

An inner window is available for the 172 to control cabin noise. The windows, similar to storm windows on a house, are installed in the doors (Fig. 11-8). By increasing the window thickness and including a dead air space, interior noise is reduced. For further information, contact:

Las Vegas Aero Specialties, Inc.
2772 North Rancho Dr.
Las Vegas, NV 89130
(702) 647-6121

Las Vegas Aero Specialties, Incorporated

Fig. 11-8. Noise-reducing interior windows.

Shoulder harnesses

Safety is a key word in aviation. Basic lap belts in automobiles and airplanes were considered safe, but automobiles lead the way with additional safety provided by shoulder harnesses. As an improvement to the Cessna 172, BAS Inc. produces an inertia reel shoulder harness to replace the single lap belt. The self-storing harness does not intrude into the cabin area and does not obstruct access to the rear seats. For additional information contact:

BAS Inc.
P.O. Box 190
Eatonville, WA 98328
(206) 832-6566

Upgraded instrumentation

Modern instrument panels no doubt have a certain allure. Review the chronological panel photographs in chapter 9 and note the progressive differences. Avion Research has designed a complete STC-PMA kit for the 172, featuring a standard T layout and center-stacked electronics (Fig. 11-9). Installation results in an updated appearance and usefulness. For further information contact:

Avion Research, Inc.
P.O. Box 597
Cupertino, CA 95014
(415) 494-7540

Fig. 11-9. Modern instrument panel including T layout, improved shock mounting, new control wheel guides and locks, rocker switches, circuit breaker holes, and internally lighted glare shield.

Door catches

Door catches on older 172s are usually rusted and no longer properly hold the doors open while on the ground. The installation of a Sky Catch will eliminate this problem. For further information, contact:

R.W. Traves Associates
829 Oak Street
Medina, OH 44256
(216) 723-2778

Fuel drain valves

A belly drain valve makes it possible to completely clear water from the fuel system. The valve takes about 10 minutes to install at the lowest point in the fuel system, where water will gather. For further information contact:

C-MODs
P.O. Box 15388
Durham, NC 27704

Oil filters

Many early aircraft engines were built without spin-on oil filters. A kit is available that fits the 172's O-300 series Continental engines and the O-320 series Lycoming engines. For further information contact:

El Reno Aviation, Inc.
P.O. Box 760
El Reno, OK 73036-0760
(405) 262-2387

Stowable seats

Rear seats of a 172 can be removed for hauling freight, equipment, or other cargo, which is fine until you need the seats for passengers and you have to reinstall the seats. Stowable rear seats are available (Fig. 11-10 and Fig. 11-11). For further information contact:

BAS Inc.
P.O. Box 190
Eatonville, WA 98328
(206) 832-6566

Fig. 11-10. BAS stowable seats positioned for passengers.

Fig. 11-11. BAS stowable seats folded away for cargo space.

12

Fun flying on floats

EVER WISH YOU COULD GO SOMEPLACE and really get away from it all? Well, perhaps the answer to your dreams is a floatplane. A floatplane does not have conventional landing gear, instead it is equipped with floats (Fig. 12-1 and Fig. 12-2). The fun of float flying attracts many people and the airplanes offer a means of transportation to otherwise inaccessible locations; however, realizing the capabilities of a floatplane, a pilot should know a few things before flying or buying one:

- You will need a seaplane rating on your certificate. Many flying schools around the country offer seaplane training that is completely practical, and involves only flying skills demonstrated on a checkride: no written examination. Often the price is fixed, and the rating guaranteed. Check *Trade-A-Plane* and other publications for appropriate advertising.
- You are forever limited to water flying. The Cessna 172 cannot be equipped with amphibious floats, due to weight limitations.
- Aircraft performance is reduced, which is reflected by the float-equipped 172 performance figures that follow.
- Considerably more maintenance is required for aircraft engaged in water operations. Much of this in the form of corrosion control.
- Your insurance rates will go up considerably, unless you have extensive floatplane experience with no accident history. Even then, rates will be higher for a floatplane than for a landplane due to the higher loss ratio with floatplanes, as compared to landplanes. For example, a typical ground loop in a land-based aircraft can result in several hundred dollars of damage. The damaged aircraft can often be taxied to a repair facility. The same type of mishap on the water could mean a sunken aircraft, resulting in difficult or impossible recovery, and thousands of dollars in damages.

PERFORMANCE FIGURES

Specifications for the various 172 models are listed by horsepower of the engine. If you compare these to the performance and specification figures in chapter 2, you will see

Fig. 12-1. A Cessna 172 on floats moored (anchored).

Fig. 12-2. Cessna Skyhawk "on the step."

quite a large difference in performance between landplanes and floatplanes. The difference is quite normal because the floats and rigging are heavy and extremely bulky.

145-hp version

Speed
    Max speed at sea level: 108 mph
    Cruise, 75 percent power at 6500 ft: 106 mph
Range
    Cruise, 75 percent at 6500 ft: 500 mi

                38 gallons, no reserve: 4.7 hrs/106 mph
                Cruise, 75 percent at 6500 ft: 625 mi
                48 gallons, no reserve: 5.9 hrs/106 mph
                Optimum range at 10,000 ft: 530 mi
                38 gallons, no reserve: 5.5 hrs/97 mph
                Optimum range at 10,000 ft: 670 mi
                48 gallons, no reserve: 7.0 hrs/97 mph
Rate of climb at sea level: 580 fpm
Service ceiling: 12,000 ft
Takeoff
        Water run: 1620 ft
        Over 50-ft obstacle: 2390 ft
Landing
        Water roll: 590 ft
        Over 50-ft obstacle: 1345 ft
Stall speed
        Flaps up, power off: 59 mph
        Flaps down, power off: 52 mph
Baggage: 120 lbs
Wing loading (lbs/sq ft): 12.7
Power loading (lbs/hp): 14.8
Fuel capacity
        Standard: 42 gal
        Long-range tanks: 52 gal
        Oil capacity: 8 qts
Engine
        Make/model: Continental O-300
        TBO: 1800 hrs
        Power: 145 hp
        Propeller (diameter): 80 in
Dimensions
        Wingspan: 35 ft 10 in
        Wing area (sq ft): 174
        Length: 27 ft 00 in
        Height: 9 ft 11 in
Weight
        Gross weight: 2220 lbs
        Empty weight: 1415 lbs
        Useful load: 805 lbs

150-hp version

Speed
        Max speed at sea level: 98 kts
        Cruise, 75 percent power at 7500 ft: 97 kts
Range
        Cruise, 75 percent at 7500 ft: 365 nm

38 gallons usable fuel: 3.8 hrs
Cruise, 75 percent at 7500 ft: 480 nm
48 gallons usable fuel: 5.0 hrs
Maximum range at 10,000 ft: 385 nm
38 gallons usable fuel: 4.4 hrs
Maximum range at 10,000 ft: 510 nm
48 gallons usable fuel: 5.9 hrs
Rate of climb at sea level: 715 fpm
Service ceiling: 12,000 ft
Takeoff
Water run: 1620 ft
Over 50-ft obstacle: 2390 ft
Landing
Water roll: 590 ft
Over 50-ft obstacle: 1345 ft
Stall speed
Flaps up, power off: 46 kts
Flaps down, power off: 44 kts
Baggage: 120 lbs
Wing loading (lbs/sq ft): 12.7
Power loading (lbs/hp): 14.8
Fuel capacity
Standard: 42 gal
Long-range tanks: 52 gal
Oil capacity: 8 qts
Engine
Make/model: Lycoming O-320-E2D
TBO: 2000 hrs
Power: 150 hp
Propeller (diameter): 80 in
Dimensions
Wingspan: 35 ft 10 in
Wing area (sq ft): 174
Length: 27 ft 00 in
Height: 9 ft 11 in
Weight
Gross weight: 2220 lbs
Empty weight: 1574 lbs
Useful load: 646 lbs

160-hp version

Speed
Max speed at sea level: 96 kts
Cruise, 75 percent power at 4000 ft: 95 kts
Range
Cruise, 75 percent power at 4000 ft: 360 nm

40 gallons usable fuel: 3.8 hrs
Cruise, 75 percent power at 4000 ft: 475 nm
50 gallons usable fuel: 5.0 hrs
Maximum range at 10,000 ft: 435 nm
40 gallons usable fuel: 5.6 hrs
Maximum range at 10,000 ft: 565 nm
50 gallons usable fuel: 7.3 hrs
Rate of climb at sea level: 740 fpm
Service ceiling: 15,000 ft
Takeoff
    Water run: 1400 ft
    Over 50-ft obstacle: 2160 ft
Landing
    Water roll: 590 ft
    Over 50-ft obstacle: 1345 ft
Stall speed
    Flaps up, power off: 48 kts
    Flaps down, power off: 44 kts
Baggage: 120 lbs
Wing loading (lbs/sq ft): 12.7
Power loading (lbs/hp): 13.9
Fuel capacity
    Standard: 43 gal
    Long-range tanks: 54 gal
    Oil capacity: 8 qts
Engine
    Make/model: Lycoming O-320-D2J
    TBO: 2000 hrs
    Power: 160 hp
    Propeller (diameter): 80 in
Dimensions
    Wingspan: 35 ft 10 in
    Wing area (sq ft): 174
    Length: 26 ft 08 in
    Height: 11 ft 11 in
Weight
    Gross weight: 2220 lbs
    Empty weight: 1621 lbs
    Useful load: 606 lbs

195-hp version

Speed
    Max speed at sea level: 118 kts
    Cruise, 75 percent power at 6000 ft: 116 kts
Range
    Cruise, 80 percent power at 6000 ft: 395 nm

49 gallons usable fuel: 3.4 hrs
Cruise, 80 percent power at 6000 ft: 570 nm
66 gallons usable fuel: 4.9 hrs
Maximum range at 10,000 ft: 495 nm
49 gallons usable fuel: 5.5 hrs
Maximum range at 10,000 ft: 705 nm
66 gallons usable fuel: 7.9 hrs
Rate of climb at sea level: 870 fpm
Service ceiling: 15,500 ft
Takeoff
    Water run: 1135 ft
    Over 50-ft obstacle: 1850 ft
Landing
    Water roll: 675 ft
    Over 50-ft obstacle: 1390 ft
Stall speed
    Flaps up, power off: 50 kts
    Flaps down, power off: 44 kts
Baggage: 200 lbs
Wing loading (lbs/sq ft): 14.7
Power loading (lbs/hp): 13.1
Fuel capacity
    Standard: 52 gal
    Long-range tanks: 68 gal
    Oil capacity: 9 qts
Engine
    Make/model: Teledyne-Cont. IO-360-KB
    TBO: 2000 hrs
    Power: 195 hp
    Propeller: (diameter) C/S 80 in
Dimensions
    Wingspan: 35 ft 10 in
    Wing area (sq ft): 174
    Length: 26 ft 08 in
    Height: 12 ft 05 in
Weight
    Gross weight: 2550 lbs
    Empty weight: 1808 lbs
    Useful load: 750 lbs

FLOATS

The following floats are available for Cessna 172 airplanes:

For the 172: Aqua 2200 and 2400.
For the 172 XP: Aqua 2400.
Aqua Floats

P.O. Box 247
Brandon, MN 56315
(612) 524-2782

For the 172: Canadian model 2000, available from:
Canadian Aircraft Products, Ltd.
2611 Viscount Way
Richmond, BC V6V1M9
Canada
(604) 278-9821

For the 172 and for the 172 XP, PK B2300:
DeVore Aviation Corp.
6104B Kircher Blvd., NE
Albuquerque, NM 87109
(505) 345-8713

For the 172: Edo 2000, 2130, and 244OB.
For the 172 XP: Edo 244OB.
Edo Corp.
Float Operations
14-04 111 Street
College Point, NY 11356
(212) 445-6000

Prices

Floats represent a major investment; however, floats tend to hold their value. The following list reflects the total of the suggested retail price plus installation estimates:

- Canadian 2000 $13,000
- Aqua 2200 $16,700
- Aqua 2400 $17,900
- Edo 2000 $16,800
- Edo 244OB $18,500
- PK (DeVore) B2300 $17,900

MODIFICATIONS

Precious few 172s on floats are available in the used market. It is possible to convert an existing 172 for use on floats; the conversion can be done in the field because no extensive disassembly is necessary. Conversion kits are available from Cessna and consist of the necessary materials to strengthen the airframe and make attachment points for the floats. Stainless steel control cables are a part of the kit. The modification requires application of zinc chromate inside, outside, and between all panels of the airframe.

The 172 XP with the 195-hp engine makes a fine floatplane, but it is now out of production. This is unfortunate because the 172 XP is a true four-place floatplane. The basic 172 is really considered to be a two-place "plus" airplane, due to the weight and horsepower limitations. Horsepower can often be increased with the installation of a larger engine, as described in chapter 11.

FLOATPLANE ACTIVITY

Floatplane flying is in every state of the union; however, some states have more activity than others. The following states offer considerable float flying fun and adventure: Alaska, California, Florida, Louisiana, Maine, Massachusetts, Minnesota, Washington, and Wisconsin. Canada offers some of the best floatplane flying in the world; many Canadian locations are accessible only by floatplane.

Waterway restrictions

Unfortunately, not all places you might want to land in are available for seaplane use. According to the Wilderness Act of 1964, the U.S. Forest Service can determine if you may fly into their waterways. Other areas, including national parks, forests, and the like, that are not designated as wilderness areas might also be on the list of restricted flying activities.

To make things even more difficult, areas will change designation with little or no notice to the user. Additionally, local and state governments can regulate locations for water flight.

A copy of the *Water Landing Directory* from the Seaplane Pilots Association in Frederick, Maryland, will keep you abreast of where you may operate. Contact SPA at the address or phone at the end of this chapter.

SEAPLANE PILOTS ASSOCIATION

AOPA sponsors the Seaplane Pilots Association, which was formed in 1972 and claims several thousand members. SPA's objective is to assist seaplane pilots with technical problems, provide a national lobbying effort, and explore possible economy measures.

Membership features the quarterly magazine *Water Flying*, the *Water Flying Annual*, and other written communications that include timely tips and safety suggestions. SPA sponsors numerous annual fly-ins. For further information, contact:

Seaplane Pilots Association
421 Aviation Way
Frederick, MD 21701
(301) 695-2000

13

Hangar flying

WORD-OF-MOUTH INFORMATION is probably the best way to fine-tune a pilot's understanding of an airplane. Asking owners and other pilots about the idiosyncrasies of a 172 can yield invaluable comments, statements, rumors, and facts about the airplane. Specific hangar flying regarding the 172 is the basis of this chapter. Ideally, you will go beyond these issues and have your own questions to ask.

CLUBS

The most information about an airplane is probably available from a national or international club that supports that airplane. Two clubs support Cessna single-engine airplanes.

Cessna Pilots Association

The Cessna Pilots Association's sole purpose is to support owners and pilots of Cessna single-engine airplanes. The association publishes a monthly slick magazine with color photos and a large amount of well-documented information pertinent to the safe and cost-effective ownership of a Cessna airplane. The magazine contains commercial advertising applicable to Cessna single-engine airplanes. The association maintains a large technical library, which is very handy when reference material is required. Membership is $40 annually. For further information contact:

Cessna Pilots Association
Mid-Continent Airport, P.O. Box 12948
Wichita, KS 67277
(800) 852-2272
(316) 946-4777
FAX (316) 943-5546

Cessna Owner Organization

The Cessna Owner Organization offers a slick color monthly magazine that includes feature articles, maintenance and modification ideas, technical information, Cessna service letters, service difficulty reports, airworthiness directives and alerts, and classified advertisements. The association has a group aircraft insurance program, a family-oriented annual fly-in, and telephone assistance. Annual dues are $38. For more information, contact:

Cessna Owner Organization
P.O. Box 337
Iola, WI 54945
(715) 445-5000
FAX (715) 445-4053

WHO SAYS WHAT

The Cessna 172 is a very talked-about airplane. After all, a lot are around. Here is a synopsis of many comments I have heard regarding the 172.

Pilots

The 172 flies just like a 150, just bigger.

It's no tiger on takeoff in the summer, but it gets there, even when we (four) are all aboard.

I've never attained the published cruise speed in my '65, but it still gets me there.

The nosewheel shimmies, just like the one on my old 150.

We bought the 172 instead of a PA-28/140 to have the capability to really carry four people.

I fly from a mountain airport at the 7,000-foot line. Don't think I would ever try to get out with four on board in the summer.

My wife gave me a set of sheepskin seat covers for Christmas last year—real class, in a class bird.

I just had a 180-hp engine installed, wow! What get-up-and-go. This really helps, as the ranch strip is kinda bad in the spring.

Most of the time I fly by myself, but I prefer the extra weight of the 172 over the lighter 150. This is very true when I am flying IFR.

The visibility in a busy traffic pattern is poor, but that's the same for all high-wingers.

Love the barn door flaps; can really save a high approach.

I have a '63 model with manual flaps, and plan to keep it. I don't like the electric flaps; there's too much that can break on them.

A faster airplane would be nice, but most of my trips are less than 300 miles, and the actual time saved would be small, but the increased maintenance would be large.

My wife is learning to fly, and when my son gets old enough, he will, too. Flying our Hawk has become a family affair.

I have the Deemers tips on the plane, and they allow me to make it into my farm strip easier. The strip is only 990 feet long, but clear on both ends.

It's too noisy, but most small planes are.

There's only the two of us, so we can take all the baggage and other stuff we want to. Loads of room.

After ground-looping my 180 a couple of times, I sold it and got this 172 with the 180-hp STC. It is really super here on the ranch; I use it like a station wagon.

I use mogas in my '64 Hawk. Seems okay, and it saves me money. Just wish the FBO would pump it; the gas in the trunk of the car scares me.

My '58 has better forward visibility than the newer models with all the gadgets in the panel.

Owners

Maintenance is what I like about my Hawk. It requires very little.

We recently traded down to a 172 from a '65 Beech 35. The service it required was breaking me. Now we can afford to fly again—slower, but we get there.

I just replaced the last of six cylinders on my Continental O-300. That's at 150 SMOH. Not all overhauls are created equal.

My last annual cost $375, and that's in the high-cost Washington, D.C. area. I'm real happy with that.

My bird has King avionics and I love them. I've never had a lick of trouble with them. My brother has ARC and is jealous. His is in the shop all the time.

Had to replace several landing lights before I heard about the vibration kit available from Cessna. Why don't people (FBOs and dealers) tell you about these things?

I could afford to buy a retractable if I wanted it, but why pay for all the extra maintenance for a little extra speed? Slow down and enjoy life.

Not a real thrill to fly, but dependable and rugged.

Take care of the paint job, they cost too !!!! much.

I just can't get over the value going up each year.

An airplane that you buy, fly, and later sell for a profit.

Mechanics

There's not much that can go wrong with a Cessna high wing that creates any mystery. They were built tough.

The four-cylinder Lycoming 150-hp engines seem to be very solid. The Continental O-300 engines develop cracks in the jugs, and the H2AD has been real headaches for all concerned.

The O-300 engine is more expensive to overhaul then the O-320.

Get the 100LL valve kit for the O-300; it really helps.

The H engines from S/N 7976 and up are modified, and the factory considers them to be the answer.

These planes are simple enough for the typical owner to care for with little supervision.

Poor fuel sump draining is a real problem, but it can be fixed for less than $50 by installing a belly drain.

The seat attach points on the seat rails are a weak point.

Watch for filiform corrosion on the 1977 through 82 models. They sometimes suffer from poor prepaint preparation at the factory. Also, on land-plane models there was never any corrosion proofing on interior surfaces.

Although the 172 is basically a very good airplane, you must remember that age is creeping up on them. The early models are fast approaching 40.

Line service personnel

I see all kinds of planes, and some are real expensive, but I really like to see an old 172 that has been all fixed up. The pilot is usually the guy who fixed it, and he's real proud of it.

High wings are harder to fuel than low wings.

Wish all Cessnas had steps on them.

All the old Cessna drivers were sure happy to see red (80/87 octane) fuel again.

Some of the 172 drivers are refueling out of their cars. Guess it saves money, but it doesn't guarantee I'll be employed much longer.

We now have a Mogas pump and a good percentage of the older 172 drivers are using it.

Sales representatives

I like to sell 172s to first-time buyers because they make people happy, not broke.

172s sell themselves. They are roomy, look good, and are reasonable in price. They're just real good value . . . something you don't often see these days.

I have rarely seen a bargain 172; generally, you get what you pay for.

There is such a 172 market that the prices are almost fixed.

I wouldn't recommend the 172 models with the Lycoming H engine—too many expensive problems.

A 172 makes a good investment. If you keep it in good shape, you'll always get your money out of it.

As a purchasing broker, I look at a lot of 172s for my customers. I am always able to find a plane to fill the bill. There are just so many around.

172s have been discovered, and there are few bargains, however, they represent an investment capable of a positive return.

It is getting harder and harder to find really good ones for sale.

I keep hearing rumors about the best planes being sent to foreign countries.

Insurance carriers

Insuring the Cessna 172 is easy, as there are no real secrets to them. They are reliable, easy to fly, and replacement parts are available everywhere.

We feel very comfortable about insuring low-time pilots in the 172.

The 172 makes flying a family affair, and that generally makes for safer flying.

The 172 airplanes are quite forgiving of pilot inattention, therefore provide a low risk for insurance coverage.

The owner is warned to update the insurance loss limits annually, as the 172 keeps increasing in value.

NTSB

The following tables of comparative accident data are a compilation of a study made by the National Transportation Safety Board (fatal accident rate comparison by manufacturer). All figures are based upon the adjusted rate of 100,000 hours of flying time. Note where the various 172 models are on the charts. The worst record is at the top and the best record is at the bottom of each category.

Accident rate
(per 100,000 hours)

Bellanca	4.84
Grumman	4.13
Beech	2.54
Mooney	2.50
Piper	2.48
Cessna	1.65

Engine failures
(per 100,000 hours)

Globe GC-1	12.36
Stinson 108	10.65
Ercoupe	9.50
Grumman AA-1	8.71
Navion	7.84
Piper J-3	7.61
Luscombe 8	7.58
Cessna 120/140	6.73
Piper PA-12	6.54
Bellanca 14-19	5.98
Piper PA-22	5.67
Cessna 195	4.69
Piper PA-32	4.39
Cessna 210/205	4.25
Aeronca 7	4.23
Aeronca 11	4.10
Taylorcraft BC	3.81

Piper PA-24	3.61
Beech 23	3.58
Cessna 175	3.48
Mooney M-20	3.42
Piper PA-18	3.37
Cessna 177	3.33
Cessna 206	3.30
Cessna 180	3.24
Cessna 170	2.88
Cessna 185	2.73
Cessna 150	2.48
Piper PA-28	2.37
Beech 33,35,36	2.22
Grumman AA-5	2.20
Cessna 182	2.08
Cessna 172	1.41

In-flight airframe failures
(per 100,000 hours)

Bellanca 14-19	1.49
Globe GC-1	1.03
Ercoupe	0.97
Cessna 195	0.94
Navion	0.90
Aeronca 11	0.59
Beech 33,35,36	0.58
Luscombe 8	0.54
Piper PA-24	0.42
Cessna 170	0.36
Cessna 210/205	0.34
Cessna 180	0.31
Piper PA-22	0.30
Aeronca 7	0.27
Beech 23	0.27
Cessna 120/140	0.27
Piper PA-32	0.24
Taylorcraft BC	0.24
Piper J-3	0.23
Mooney M-20	0.18
Piper PA-28	0.16
Cessna 177	0.16
Cessna 182	0.12
Cessna 206	0.11
Grumman AA-1	0.09
Cessna 172	0.03
Cessna 150	0.02

Stalls
(per 100,000 hours)

Aeronca 7	22.47
Aeronca 11	8.21
Taylorcraft BC	6.44
Piper J-3	5.88
Luscombe 8	5.78
Piper PA-18	5.49
Globe GC-1	5.15
Cessna 170	4.38
Grumman AA-1	4.23
Piper PA-12	3.27
Cessna 120/140	2.51
Stinson 108	2.09
Navion	1.81
Piper PA-22	1.78
Cessna 177	1.77
Grumman AA-5	1.76
Cessna 185	1.47
Cessna 150	1.42
Beech 23	1.41
Ercoupe	1.29
Cessna 180	1.08
Piper PA-24	0.98
Beech 33,35,36	0.94
Cessna 175	0.83
Piper PA-28	0.80
Mooney M-20	0.80
Cessna 172	0.77
Cessna 210/205	0.71
Bellanca 14-19	0.60
Piper PA-32	0.57
Cessna 206	0.54
Cessna 195	0.47
Cessna 182	0.36

Hard landings
(per 100,000 hours)

Beech 23	3.50
Grumman AA-1	3.02
Ercoupe	2.90
Cessna 177	2.60
Globe GC-1	2.58
Luscombe 8	2.35
Cessna 182	2.17
Cessna 170	1.89

Beech 33,35,36	1.45
Cessna 150	1.37
Cessna 120/140	1.35
Cessna 206	1.30
Piper PA-24	1.29
Aeronca 7	1.20
Piper J-3	1.04
Grumman AA-5	1.03
Cessna 175	1.00
Cessna 180	0.93
Cessna 210/205	0.82
Piper PA-28	0.81
Cessna 172	0.71
Piper PA-22	0.69
Taylorcraft BC	0.48
Cessna 195	0.47
Piper PA-18	0.43
Piper PA-32	0.42
Cessna 185	0.42
Navion	0.36
Mooney M-20	0.31
Piper PA-12	0.23
Stinson 108	0.19

Ground loops
(per 100,000 hours)

Cessna 195	22.06
Stinson 108	13.50
Luscombe 8	13.00
Cessna 170	9.91
Cessna 120/140	8.99
Aeronca 11	7.86
Aeronca 7	7.48
Cessna 180	6.49
Cessna 185	4.72
Piper PA-12	4.67
Piper PA-18	3.90
Taylorcraft BC	3.58
Globe GC-1	3.09
Grumman AA-1	2.85
Piper PA-22	2.76
Ercoupe	2.74
Beech 23	2.33
Bellanca 14-19	2.10
Piper J-3	2.07
Cessna 206	1.73
Cessna 177	1.61
Grumman AA-5	1.47

Piper PA-32	1.42
Cessna 150	1.37
Piper PA-28	1.36
Piper PA-24	1.29
Cessna 210/205	1.08
Cessna 182	1.06
Cessna 172	1.00
Mooney M-20	0.65
Beech 33,35,36	0.55
Navion	0.36
Cessna 175	0.17

**Landing undershoots
(per 100,000 hours)**

Ercoupe	2.41
Luscombe 8	1.62
Piper PA-12	1.40
Globe GC-1	1.03
Cessna 175	0.99
Grumman AA-1	0.95
Taylorcraft BC	0.95
Piper PA-22	0.83
Piper PA-32	0.70
Bellanca 14-19	0.60
Aeronca 11	0.59
Piper PA-28	0.59
Aeronca 7	0.59
Piper PA-24	0.57
Piper J-3	0.57
Stinson 108	0.57
Cessna 120/140	0.53
Cessna 195	0.47
Grumman AA-5	0.44
Piper PA-18	0.43
Beech 23	0.43
Cessna 185	0.41
Mooney M-20	0.37
Cessna 170	0.36
Navion	0.36
Cessna 150	0.35
Cessna 210/205	0.33
Cessna 206	0.32
Cessna 172	0.26
Cessna 182	0.24
Beech 33,35,36	0.21
Cessna 180	0.15
Cessna 177	0.10

**Landing overshoots
(per 100,000 hours)**

Grumman AA-5	2.35
Cessna 195	2.34
Beech 23	1.95
Piper PA-24	1.61
Piper PA-22	1.33
Cessna 175	1.33
Stinson 108	1.33
Cessna 182	1.21
Aeronca 11	1.17
Luscombe 8	1.08
Piper PA-32	1.03
Globe GC-1	1.03
Mooney M-20	1.01
Cessna 172	1.00
Cessna 170	0.99
Grumman AA-1	0.95
Piper PA-12	0.93
Cessna 210/205	0.89
Cessna 177	0.88
Piper PA-18	0.81
Cessna 206	0.81
Piper PA-28	0.80
Cessna 120/140	0.71
Ercoupe	0.64
Bellanca 14-19	0.60
Cessna 180	0.56
Navion	0.54
Aeronca 7	0.48
Cessna 150	0.35
Piper J-3	0.34
Cessna 185	0.31
Beech 33,35,36	0.23

TIPS

Never leave a 172 unattended and not tied down. It could not only move in the wind and get damaged, it could damage someone else's airplane.

Plug the air intakes of the cowling to keep birds out.

Use pitot tube covers to keep bugs from blocking the inlet.

Mount a fire extinguisher in the cabin where you can quickly reach it.

Keep a working flashlight on board, better yet, why not two flashlights?

Always carry a complete first-aid kit in the 172.

If flying over sparsely populated areas, it would be a good idea to have survival water, food, a tarp or tent, and blankets or sleeping bags on board.

Flares would be handy when trying to attract attention during rescue operations.

Appendix A

Advertising abbreviations

AD airworthiness directive
ADF automatic direction finder
AF airframe
AF&E airframe and engine
AI aircraft inspector
ALT altimeter
ANN annual inspection
AP autopilot
ASI airspeed indicator
A&E airframe and engine
A/P autopilot
BAT battery
CAT carburetor air temperature
CHT cylinder head temperature
COM communications radio
COMM communications radio
CS constant-speed propeller
C/S constant-speed propeller
C/W complied with
DBL double
DG directional gyro
FAC factory
FBO fixed-base operator
FGP full gyro panel
FWF firewall forward
G gravity
GAL gallons
GPH gallons per hour
GS glideslope

HD heavy duty
HP horsepower
IFR instrument flight rules
INSP inspection
INST instrument
KTS knots
L left
LDG landing
LED light emitting diode
LH left-hand
LIC license
LOC localizer
LTS lights
L&R left and right
MB marker beacon
MBR marker beacon receiver
MP manifold pressure
MPH miles per hour
MOD modification
NAV navigation
NAV/COM navigation/communication radio
NAV/COMM navigation/communication radio
NAVCOM navigation/communication radio
NDH no damage history
OAT outside air temperature
PMA parts manufacture approval
PROP propeller
R right
RC rate of climb
REMAN remanufactured
RH right-hand
RMFD remanufactured
RMFG remanufactured
ROC rate of climb
SAFOH since airframe overhaul
SCMOH since chrome major overhaul
SEL single engine land
SFACNEW since factory new
SFN since factory new
SFNE since factory new engine
SFREM since factory remanufacture
SFREMAN since factory remanufacture
SFRMFG since factory remanufacture
SMOH since major overhaul
SN serial number
SNEW since new
SPOH since propeller overhaul
STC supplemental type certificate

STOH since top overhaul
STOL short takeoff and landing
TAS true airspeed
TBO time between overhaul
TC turbocharged
TNSP transponder
TNSPNDR transponder
TSN time since new
TSO technical service order
TT total time
TTAF total time airframe
TTA&E total time airframe and engine
TTE total time engine
TTSN total time since new
TXP transponder
T&B turn and bank
VAC vacuum
VFR visual flight rules
VHF very high frequency
VOR VHF omnidirectional range
XC cross-country
XMTR transmitter
XPDR transponder
XPNDR transponder

Appendix B

Telephone area codes

201 New Jersey north (Newark)
202 District of Columbia (Washington)
203 Connecticut
204 Manitoba, Canada
205 Alabama
206 Washington west (Seattle)
207 Maine
208 Idaho
209 California central (Fresno)
212 New York southeast (New York City)
213 California southwest (Los Angeles)
214 Texas northeast (Dallas)
215 Pennsylvania southeast (Philadelphia)
216 Ohio northeast (Cleveland)
217 Illinois central (Springfield)
218 Minnesota north (Duluth)
219 Indiana north (South Bend)

301 Maryland
302 Delaware
303 Colorado
304 West Virginia
305 Florida southeast (Miami)
306 Saskatchewan, Canada
307 Wyoming
308 Nebraska west (North Platte)
309 Illinois northwest (Peoria)
310 California southwest (Los Angeles)
312 Illinois northeast (Chicago)
313 Michigan southeast (Detroit)
314 Missouri east (St. Louis)
315 New York north central (Syracuse)

316 Kansas south (Wichita)
317 Indiana central (Indianapolis)
318 Louisiana northwest (Shreveport)
319 Iowa east (Dubuque)

401 Rhode Island
402 Nebraska east (Omaha)
403 Alberta, Canada
404 Georgia north (Atlanta)
405 Oklahoma west (Oklahoma City)
406 Montana
407 Florida east (Melbourne)
408 California northwest (San Jose)
409 Texas southeast (Galveston)
410 Maryland eastern shore
412 Pennsylvania southwest (Pittsburgh)
413 Massachusetts west (Springfield)
414 Wisconsin southeast (Milwaukee)
415 California central (San Francisco)
416 Ontario, Canada
417 Missouri southwest (Springfield)
418 Quebec, Canada
419 Ohio northwest (Toledo)

501 Arkansas
502 Kentucky west (Louisville)
503 Oregon
504 Louisiana southeast (New Orleans)
505 New Mexico
506 New Brunswick, Canada
507 Minnesota south (Rochester)
508 Massachusetts east (except Boston)
509 Washington east (Spokane)
510 California central (Oakland)
512 Texas south (San Antonio)
513 Ohio southwest (Cincinnati)
514 Quebec, Canada
515 Iowa central (Des Moines)
516 New York southeast (Long Island)
517 Michigan central (Lansing)
518 New York northeast (Albany)
519 Ontario, Canada

601 Mississippi
602 Arizona
603 New Hampshire
604 British Columbia, Canada
605 South Dakota

606 Kentucky east (Lexington)
607 New York south central (Binghamton)
608 Wisconsin southwest (Madison)
609 New Jersey south (Trenton)
612 Minnesota central (Minneapolis)
613 Ontario, Canada
614 Ohio southeast (Columbus)
615 Tennessee east (Nashville)
616 Michigan west (Grand Rapids)
617 Massachusetts east (Boston)
618 Illinois south (Centralia)
619 California south (San Diego)

701 North Dakota
702 Nevada
703 Virginia north and west (Arlington)
704 North Carolina west (Charlotte)
705 Ontario, Canada
707 California northwest (Santa Rosa)
708 Illinois north (Chicago)
709 Newfoundland, Canada
712 Iowa west (Council Bluffs)
713 Texas southeast (Houston)
714 California southwest (Orange)
715 Wisconsin north (Eau Claire)
716 New York west (Buffalo)
717 Pennsylvania central (Harrisburg)
718 New York southeast (Brooklyn)
719 Colorado south (Pueblo)

801 Utah
802 Vermont
803 South Carolina
804 Virginia southeast (Richmond)
805 California west central (Bakersfield)
806 Texas northwest (Amarillo)
807 Ontario, Canada
808 Hawaii
809 Bermuda, Puerto Rico, Virgin Islands, and other islands
812 Indiana south (Evansville)
813 Florida southwest (Tampa)
814 Pennsylvania northwest (Altoona)
815 Illinois northwest (Rockford)
816 Missouri northwest (Kansas City)
817 Texas north central (Fort Worth)
818 California southwest (Pasadena)
819 Quebec, Canada

901 Tennessee west (Memphis)
902 Nova Scotia, Canada
903 Mexico
904 Florida north (Jacksonville)
905 Mexico City, Mexico
906 Michigan northwest (Escanaba)
907 Alaska
908 New Jersey central
912 Georgia south (Savannah)
913 Kansas north (Topeka)
914 New York southeast (White Plains)
915 Texas southwest (San Angelo)
916 California northwest (Sacramento)
918 Oklahoma northeast (Tulsa)
919 North Carolina east (Raleigh)

Appendix C

FAA office locations

Whenever you need a question answered, you can always turn to the Federal Aviation Administration. The FAA has many offices spread around the country with experts to serve the public. Remember, FAA employees are government employees, there to serve, and paid with tax dollars. In the many years that I have been associated with general aviation, I have never been disappointed with the help I received from the FAA. When a problem arises, contact them.

Alabama

FSDO 09
6500 43rd Avenue North, Birmingham, AL 35206
(205) 731-1393

Alaska

FSDO 01
6348 Old Airport Way, Fairbanks, AK 99709
(907) 474-0276

FSDO 03
4510 West Int'l Airport Road, Suite 302, Anchorage, AK 99502-1088
(907) 243-1902

FSDO 05
1910 Alex Holden Way, Suite A, Juneau, AK 99801
(907) 789-0231

Arizona

FSDO
15041 North Airport Drive, Scottsdale, AZ 85260
(602) 640-2561

Arkansas

FSDO 11
1701 Bond Street, Little Rock, AR 72202
(501) 324-5565

California

LAX FSDO
5885 West Imperial Hwy, Los Angeles, CA 90045
(310) 215-2150

FSDO
831 Mitten Rd., Rm 105, Burlingame, CA 94010
(415) 876-2771

FSDO
16501 Sherman Way, Suite 330, Van Nuys, CA 91406
(818) 904-6291

FSDO
1250 Aviation Ave., Suite 295, San Jose, CA 95110-1130
(408) 291-7681

FSDO
8525 Gibbs Drive, Suite 120, San Diego, CA 92123
(619) 557-5281

FSDO
Fresno Air Terminal
4955 East Anderson, Suite 110, Fresno, CA 93727-1521
(209) 487-5306

FSDO
6961 Flight Road, Riverside, CA 92504
(714) 276-6701

FSDO
6650 Belleau Wood Lane, Sacramento, CA 95822
(916) 551-1721

FSDO
P.O. Box 2397, Airport Station, Oakland, CA 94614
(510) 273-7155

FSDO
2815 East Spring Street, Long Beach, CA 90806
(310) 426-7134

Colorado

FSDO
5440 Roslyn St., Suite 201, Denver, CO 80216
(303) 286-5400

Connecticut

FSDO 03
Building 85-214 1st Flr, Bradley International Airport, Windsor Locks, CT 06096
(203) 654-1000

Florida

FSDO
FAA Building, Craig Municipal Airport, 855 Saint John's Bluff Road, Jacksonville, FL
  32225
(904) 641-7311

FSDO 15
9677 Tradeport Dr., Suite 100, Orlando, FL 32827
(407) 648-6840

FSDO 17
286 South West 34th St., Ft. Lauderdale, FL 33315
(305) 463-4841

FSDO 19
P.O. Box 592015, Miami, FL 33159
(305) 526-2572

Georgia

FSDO 11
1680 Phoenix Pkwy., Second Flr, College Park, GA 30349
(404) 994-5276

Hawaii

FSDO
90 Nakolo Place, Room 215, Honolulu, HI 96819
(808) 836-0615

Illinois

FSDO 03
Post Office Box H, DuPage County Airport, West Chicago, IL 60185
(708) 377-4500

FSDO 19
Capitol Airport, #3 North Airport Dr., Springfield, IL 62708
(217) 492-4238

FSDO 31
9950 West Lawrence Ave., Suite 400, Schiller Park, IL 60176
(312) 353-7787

Indiana

FSDO 11
6801 Pierson Drive, Indianapolis, IN 46241
(317) 247-2491

FSDO 17
1843 Commerce Drive, South Bend, IN 46628
(219) 236-8480

Iowa

FSDO 01
3021 Army Post Road, Des Moines, IA 50321
(515) 285-9895

Kansas

FSDO 64
Mid-Continent Airport, 1801 Airport Rd., Rm 103, Wichita, KS 67209
(316) 941-1200

Kentucky

FSDO
Kaden Building, 5th Flr, 6100 Dutchmans Lane, Louisville, KY 40205-3284
(502) 582-5941

Louisiana

FSDO
Ryan Airport, 9191 Plank Rd., Baton Rouge, LA 70811
(504) 356-5701

Maine

FSDO 05
2 Al McKay Ave., Portland, ME 04102
(207) 780-3263

Maryland

FSDO 07
P.O. Box 8747, BWI Airport, MD 21240
(410) 787-0040

Massachusetts

FSDO 01
Civil Air Terminal Bldg., 2nd Flr., Hanscom Field, Bedford, MA 01730
(617) 274-7130

Michigan

FSDO 09
Kent County International Airport, P.O. Box 888879, Grand Rapids, MI 49588-8879
(616) 456-2427

FSDO 23
Willow Run Airport (east side), 8800 Beck Road, Belleville, MI 48111
(313) 487-7222

Minnesota

FSDO 15
6020 28th Ave. South, Rm 201, Minneapolis, MN 55450
(612) 725-4211

Mississippi

FSDO 07
120 North Hangar Drive, Suite C, Jackson Municipal Airport, Jackson, MS 39208
(601) 965-4633

Missouri

FSDO 03
FAA Bldg, 10801 Pear Tree Ln., Suite 200, St. Ann, MO 63074
(314) 429-1006

FSDO 05
Kansas City International Airport, 525 Mexico City Ave., Kansas City, MO 64153
(816) 243-3800

Montana

FSDO
FAA Building, Room 3, Helena Airport, Helena, MT 59601
(406) 449-5270

Nebraska

FSDO 09
General Aviation Building, Lincoln Municipal Airport, Lincoln, NE 68524
(402) 437-5485

Nevada

FSDO
210 South Rock Blvd., Reno, NV 89502
(702) 784-5321

FSDO
6020 South Spencer, Suite A7, Las Vegas, NV 89119
(702) 388-6482

New Jersey

FSDO 25
150 Fred Wehran Drive, Room 1, Teterboro Airport, Teterboro, NJ 07608
(201) 288-1745

New Mexico

FSDO
1601 Randolph Road, SE Suite 200N, Albuquerque, NM 87106
(505) 247-0156

New York

FSDO 01
Albany County Airport, Albany, NY 12211
(518) 869-8482

FSDO 11
Administration Building, Suite #235 RTE 110, Republic Airport, Farmingdale, NY 11735
(516) 694-5530

FSDO 15
181 South Franklin Ave, 4th Flr, Valley Stream, NY 11581
(718) 917-1848

FSDO 23
1 Airport Way, Suite 110, Rochester, NY 14624
(716) 263-5880

North Carolina

FSDO 05
8025 North Point Blvd., Suite 250, Winston-Salem, NC 27106
(919) 631-5147

FSDO 06
2000 Aerial Center Pkwy., Suite 120, Morrisville, NC 27263
(919) 840-5510

FSDO 08
FAA Building, 5318 Morris Field Drive, Charlotte, NC 28208
(704) 359-8471

North Dakota

FSDO 21
1801 23rd Ave. N., Fargo, ND 58102
(701) 232-8949

Ohio

CVG FSDO
4242 Airport Road, Lunken Executive Building, Cincinnati, OH 45226
(513) 533-8110

FSDO 07
3939 International Gateway, 2nd Flr., Port Columbus International Airport, Columbus, OH 43219
(614) 469-7476

FSDO
Federal Facilities Building, Cleveland Hopkins International Airport, Cleveland, OH 44135
(216) 265-1345

Oklahoma

FSDO
1300 South Meridian, Suite 601, Bethany, OK 73108
(405) 231-4196

Oregon

FSDO
Portland/Hillsboro Airport, 3355 NE Cornell Road, Hillsboro, OR 97124
(503) 326-2104

Pennsylvania

FSDO 03
Allegheny County Airport, Terminal Bldg., Rm 213, West Mifflin, PA 15122
(412) 462-5507

FSDO 05
3405 Airport Rd. North, Allentown, PA 18103
(215) 264-2888

FSDO 13
400 Airport Dr., Rm 101, New Cumberland, PA 17070
(717) 774-8271

FSDO 17
Scott Plaza 2, 2nd Flr., Philadelphia, PA 19113
(215) 596-0673

FSDO 19
One Thorn Run Center, Suite 200, 1187 Thorn Run Ext., Corapolis, PA 15108
(412) 644-5406

South Carolina

FSDO 13
103 Trade Zone Dr., West Columbia, SC 29169
(803) 765-5931

South Dakota

FSDO
Rapid City Regional Airport, Route 2 Box 4750, Rapid City, SD 57701
(605) 393-1359

Tennessee

FSDO 03
2 International Plaza Dr., Suite 700, Nashville, TN 37217
(615) 781-5437

FSDO 04
3385 Airways Blvd., Suite 115, Memphis, TN 38116
(901) 544-3820

Texas

FSDO
7701 North Stemmons Fwy, Suite 300, Lock Box 5, Dallas, TX 75247
(214) 767-5850

FSDO
Dallas-Fort Worth Regional Airport, P.O. Box 619020, Dallas-Ft. Worth Airport, TX 75261
(214) 574-2150

FSDO
8800 Paul B. Koonce Drive, Room 152, Houston, TX 77061-5190
(713) 640-4400

FSDO
Route 3, Box 51, Lubbock, TX 79401
(806) 762-0335

FSDO
10100 Reunion Place, Suite 200, San Antonio, TX 78216
(512) 341-4371

Utah

FSDO
116 North 2400 West, Salt Lake City, UT 84116
(801) 524-4247

Virginia

FSDO 21
Richmond International Airport, Executive Terminal, 2nd Flr., Sandstone, VA 23150-
2594
(804) 222-7494

FSDO 27
GT Building, Suite 112, Box 17325, Dulles International Airport, Washington, D.C.
20041
(703) 557-5360

Washington

FSDO
1601 Lind Ave., SW, Renton, WA 98055-4046
(206) 227-2810

West Virginia

FSDO 09
Yeager Airport, 301 Eagle Mountain Road, Room 144, Charleston, WV 25311
(304) 343-4689

Wisconsin

FSDO 13
4915 South Howell Avenue, 4th Flr., Milwaukee, WI 53207
(414) 747-5531

Appendix D

Price guide

PRICES FOR THE 172 AIRPLANES that are listed in this appendix are based upon average asking and selling prices during the winter of 1992–93. As with most aircraft, there are no hard and set prices because used aircraft values are controlled by several factors that are based upon the physical characteristics of the airplane:

- Age and general condition of the airplane
- How well-equipped the plane is and the age and condition of that equipment
- Remaining time on components with limited life
- History of the aircraft and past usage
- Any damage history

Two other points—perhaps the most important when setting the asking and selling price range—must be considered:

- How badly does the owner want to sell the airplane?
- How badly does the purchaser want to buy the airplane?

The meeting of these two points is called bargaining. Plan to do a lot of bargaining when purchasing an airplane. No price is ever set in concrete until the transaction is completed.

Realize that the asking price for a Cessna 172 manufactured before 1969 is generally three times the price of the airplane when new. Also noteworthy is the fact that 172s have increased in value nearly 50 percent since the first edition of this book was written in 1987. (Research found one dealer that would rather invest in 172s than the stock market because 172s have been a sure thing.)

The following prices are based upon an airplane with average equipment (avionics) and with middle time on the engine. Middle time on the engine is the middle one-third of the TBO (from 600–1200 hours on an engine with an 1800-hour TBO). Prices in 1987 are presented for comparison purposes. Pristine examples with comparatively few hours, no corrosion, excellent appearance, and no damage history might be priced considerably higher than this list shows.

Prices for the Skyhawk and Skyhawk II are somewhat higher than the standard 172 because the Skyhawk and Skyhawk II have more factory-installed options than the standard model; the same applies to the Skylark, Hawk, and Cutlass series. Additionally, there might be slight differences between the plain upgraded version and the II version, depending upon inclusion of the factory's NAVPAC avionics and instrumentation packages.

Year	Model	1987 price ($)	1993 price ($)
1956	172	8,300	16,900–17,500
1957	172	8,400	16,200–17,800
1958	172	8,450	16,500–18,000
1959	172	8,500	16,900–18,200
1960	172A	8,900	17,100–18,600
1961	172B	9,300	17,400–19,000
1962	172C	9,500	17,900–19,500
1963	172D	9,800	18,200–20,300
1964	172E	10,000	18,500–20,500
1965	172F	10,500	19,400–21,700
1966	172G	10,700	19,600–22,000
1967	172H	10,900	19,900–22,400
1968	172I	11,400	20,200–22,600
1969	172K	11,600	20,500–23,000
1970	172K	12,200	21,300–23,800
1971	172L	13,000	22,500–25,200
1972	172L	13,400	23,500–25,600
1973	172M	14,000	25,400–30,000
1974	172M	14,500	26,400–30,500
1975	172M	15,200	28,000–31,300
1976	172M	16,900	29,000–32,500
1977	172N	15,000	27,500–30,800
1977	R172K XP	23,500	35,000–39,200
1978	172N	16,000	29,500–32,000
1978	R172K XP	25,000	36,500–40,800
1979	172N	17,000	33,000–36,900
1979	R172K XP	26,500	40,200–45,000
1980	172N	20,000	37,000–41,400
1980	R172K XP	30,000	42,600–47,700
1980	Cutlass RG	32,000	45,600–48,900
1981	172P	28,000	43,500–48,300
1981	R172K XP	34,900	46,700–56,000
1981	Cutlass RG	44,500	51,200–56,100
1982	172P	39,000	48,500–57,200
1982	Cutlass RG	60,000	55,500–59,800
1983	172P	46,000	52,000–58,900
1983	172Q Cutlass	55,000	53,000–60,100
1983	Cutlass RG	70,000	60,000–67,300
1984	172P	n/a	58,500–65,500
1984	172Q Cutlass	n/a	59,100–66,000

Year	Model	1987 price ($)	1993 price ($)
1984	Cutlass RG	n/a	more than 76,000
1985	172P	n/a	more than 60,000
1986	172P	n/a	more than 67,000
1958	175	8,500	12,700–17,100
1959	175	8,700	13,600–17,500
1960	175A	9,000	14,800–18,000
1961	175B	9,400	15,000–18,500
1962	175C	9,600	15,900–19,000

Appendix E

Cessna 172
prepurchase checklist

THE PREPURCHASE CHECKLIST is intended to provide a means of comparing airplanes you have considered for purchase. It is a means of remembering specific details
about each airplane that you have looked at.

Year _____ Model _____ Color _____ N–number_____

The plane is located at _____

It was advertised in _____

INTERVIEW (by telephone or in person)
Why selling _____
General appearance/condition of the plane (1–10)_____
How many total hours on the airframe _____
How many hours on the engine since new _____
How many hours since the last overhaul _____
What type of overhaul was done _____
Who did the overhaul _____
Is there any damage history yes/no _____
What is the asking price _____

VISUAL INSPECTION
CABIN
Appearance (1–10) _____
Overall condition (1–10) _____
Upholstery (1–10) _____
Carpet (1–10) _____
Headliner (1–10) _____
Door panels (1–10) _____
Windshield condition (1–10) _____
Side and rear window condition (1–10) _____
Door operation (1–10) _____
Seat track condition (1–10) _____
Instrument panel appearance (1–10) _____
Condition of instruments (1–10) _____
List of avionics _____

Age of instruments/avionics _____
IFR or VFR _____
Workability (1–10) _____
Remarks _____

AIRFRAME
Appearance (1–10) _____
Overall condition (1–10) _____
Paint (1–10) _____
Visible corrosion _____
Visible rust _____
Dents, wrinkles, or tears in skin _____
Fuel leaks yes/no _____
Landing gear (1–10) _____
Tires (1–10) _____
Oleo strut (1–10) _____
Control surface movement _____
Control surface alignment _____
Control surface damage _____

ENGINE
Signs of oil leaks _____
Condition of hoses and lines (1–10) _____
Linkage or cable damage _____
Battery box condition (1–10) _____
Condition of Propeller (1–10) _____
Propeller damage _____
Exhaust system (1–10) _____
Color of exhaust pipe residue _____
Exhaust stains on the belly _____

LOGBOOKS
Required paperwork:
    Airworthiness Certificate _____
    Aircraft Registration Certificate _____
    FCC Station License _____
    Flight manual or operating limitations _____
    Logbooks (airframe, engine and propeller) _____
    Current equipment list _____
    Weight & balance chart _____
Airframe Logbook
    List of ADs and dates of compliance _____
    Total time since new _____
    Minor repairs noted _____
    Major repairs noted _____
    Total time and date of last inspection _____
    Modifications _____
    Form 337s _____
Engine Logbook
    Original engine yes/no _____
    List of ADs and dates of compliance _____
    Total time since new _____
    Total time since overhaul _____
    Who did the overhaul _____
    Minor repairs noted _____
    Major repairs noted _____
    Total time and date of last inspection _____
    Compression check report _____
    Modifications _____
    Form 337s _____
    Time between oil changes _____
    Oil analysis _____

REMARKS

THE TEST FLIGHT

Airplane insured by ——————————————

Pilot's name ——————————————

Pilot's ratings and medical status ——————————

Engine start ————————————————
Gauge action ————————————————
Run–up ————————————————
Control action ————————————————
Ventilation/heat ————————————————
Turns ————————————————
Stalls ————————————————
Level flight ————————————————
Hands–off trim ————————————————
Brake operation ————————————————
Wheel shimmy ————————————————
Avionics comments ————————————————
————————————————

REMARKS

MECHANIC'S INSPECTION

Mechanic's name ——————————————

Ratings and certificate number ————————————

Cost of inspection ——————————————

Outstanding ADs ——————————————

Compression check ——————————————

Borescope inspection ——————————————

REMARKS

NOTES AND COMMENTS

CURRENT OWNER INFORMATION

Name ——————————————

Address ——————————————

Phone number ——————————————

REMARKS

Appendix F

Suppliers

ADDRESSES AND TELEPHONE NUMBERS that follow are for products shown in the book, but not mentioned in the text.

Seat covers

Cooper Aviation
2149 E. Pratt Blvd.
Elk Grove, IL 60007
(708) 364-2600

Sheepskin Sources
600 W. Grand, Suite #205
Hot Springs, AR 71901
(800) 635-4113

Engine preheaters

E-Z Heat, Inc.
779 Lakeview Dr.
Chetet, WI 5472
(800) 468-4459

Aircraft Accessories
3920 Balchen Dr.
Anchorage, AK 99517
(800) 770-8108

Digital instruments

Davtron
427 Hillcrest Way
Redwood City, CA 94062
(415) 369-1188

Insight Instrument Corp.
Box 194 Ellicott Station
Buffalo, NY 14205-0194
(716) 852-3217

Avionics

Bendix King Radio Corp.
400 N. Rogers Rd.
Olathe, KS 66062
(913) 782-0400

Collins Avionics
400 Collins Rd NE
Cedar Rapids, IA
(319) 395-1000

David Clark Company, Inc.
360 Franklin St.
Worcester, MA 01615-0054
(508) 756-6216

IIMorrow Inc.
2345 Turner Rd SE
Salem, OR 97302
(800) 525-6726

Narco Avionics
P.O. Box 277
Laguna Beach, CA 92652
(800) 223-3636

Radio Systems Technology
12493 Loma Rica Dr.
Grass Valley, CA 95945
(916) 272-2203

Ross Engineering Company
12505 Starkey Rd.
Largo, FL 34643
(813) 536-1226

Telex Communications, Inc.
9600 Aldrich Ave. S.
Minneapolis, MN 55420
(612) 884-4051

VAL Avionics Ltd.
P.O. Box 13025
Salem, OR 97309-1025
(800) 255-1511

Voyager
9610 De Soto Ave.
Chatsworth, CA 91311
(818) 998-1216

Index